Scotland and Oil

Edited by A. MacGregor Hutcheson and Alexander Hogg

Cartography by H. Smith and K. Ward

Second edition

Please note that Figs. 2. [...] leum Ltd.
(Source: *B.P. Statistical* [...]
Also on page 28 (top of [...] (column 2,
line 2) please read 'given [...]

ver & Boyd
Edinburgh and
New York

OLIVER & BOYD
Croythorn House
23 Ravelston Terrace
Edinburgh EH4 3TJ

A Division of Longman Group Limited

Published in the United States of America
by Longman Inc., New York.

ISBN 0 05 002896 0
Second edition copyright
© A MacGregor Hutcheson and Alexander Hogg, 1975

Library of Congress Cataloging in Publication Data
Main entry under title:

Scotland and oil.

   Originally published by the Royal Scottish Geographical Society, Aberdeen, as no. 5 of Teachers' bulletin.
   Bibliography: p.
   1. Petroleum in submerged lands - - Scotland.
2. Scotland - - Economic conditions. I. Hutcheson,
Alexander MacGregor, 1930-   II. Hogg, Alexander, 1926-
TN874.G72S3J   338.2'7'28209411   75-23036
ISBN 0-05-002896-0

Text set in 10/12 pt. IBM Press Roman, printed by
photolithography, and bound in Great Britain at
The Pitman Press, Bath

Contents

Acknowledgements

In the preparation of this book the authors have received generous assistance from a very large number of organisations and individuals, to whom they wish to express their gratitude. This help has variously taken the form of access to or provision of maps, plans and up-to-date information concerning developments and current research; also much valuable advice as to how the first edition of *Scotland and Oil* could best be improved and extended. In the circumstances it would be invidious to name particular organisations and individuals. Appropriate credit has been given on maps and in chapter references. All figures have been drawn from base maps, etc., prepared by the individual contributors. Where not otherwise acknowledged, maps in Chapter 6 are based upon Ordnance Survey maps with the sanction of the Controller of Her Majesty's Stationery Office. Crown Copyright Reserved.

Contributors

ALEXANDER HOGG, M.A., Senior Lecturer in Geography, Aberdeen College of Education.

A. MACGREGOR HUTCHESON, M.A., Ph.D., F.R.G.S., Lecturer in Geography, Aberdeen College of Education.

J. RICHARD G. JENNINGS, B.A., Ph.D., Head of Department of Geography, Aberdeen College of Education.

HANCE D. SMITH, B.Sc., Ph.D., Lecturer in Geography, Newcastle upon Tyne Polytechnic.

KENNETH WARD, B.Sc., Lecturer in Geography, Aberdeen College of Education.

Preface

The first edition of *Scotland and Oil*, published in February 1974 by the Royal Scottish Geographical Society, was produced by the present contributors together with other members of the Department of Geography, Aberdeen College of Education. Its objective was to provide an up-to-date source of material for teachers and pupils in schools, but in the event its detailed coverage of the oil industry in Scotland proved of interest to the general public and to many firms and individuals engaged in the industry itself.

Since the book was first written events have moved quickly. New discoveries, the expansion in Scotland of supporting servicing and manufacturing industry, and major changes in the world oil supply situation have made necessary the almost complete re-writing of the first edition and the production of a new set of maps and diagrams.

As in the first edition, chapters can be read in almost any order, as cross-references are supplied. The mode of origin of chapters differs considerably. Chapter 3, for example, is essentially a review article based on a wide range of published and unpublished sources, while Chapter 6 is heavily dependent on field mapping and detailed plans and information provided by developers and local and port authorities.

Information is correct to the end of January 1975, and the bibliography has been widened to include a range of the more authoritative oil industry journals. To facilitate comparison regional maps have been drawn to a limited range of scales. The map of Shetland, is for instance, on the same scale as that of Tayside and the Firth of Forth. Symbols have similarly been standardised wherever possible.

A. M. Hutcheson
A. Hogg

Aberdeen, February 1975.

1 Introduction—Scotland and Oil

'No branch of industry has risen more rapidly than the manufacture of oils.'
D. Bremner, *Industries of Scotland,* 1869

The processing of oil from shale and coal to which Bremner refers prepared the way for the present world oil industry. However, the oil processing pioneered by James 'Paraffin' Young in central Scotland in the 1850s, like many of the regional economies of the time, existed in relative isolation. By contrast the new oil industry has to be viewed on a world scale, and especially so in relation to trading patterns. For the first time Western Europe has the opportunity of meeting an appreciable proportion of its petroleum requirements from indigenous resources, while the United States, formerly self-sufficient, has become a net importer. Further, the recent steep rise in the world price of oil and restrictions placed on oil production by some of the larger Arab producers have increased the interest of the Western World in developing Europe's offshore resources.

The exploration of these resources and associated developments onshore provide the greatest potential for change in Scotland since the advent of the first Industrial Revolution. The processes of economic development which then began brought into being regional economies based upon distinctive types of production which subsequently, in the twentieth century, have been replaced by larger economic regions centred on big towns and cities. The reality of these city regions today is a major reason behind the present re-organisation of local government. Meanwhile, at the national and international scale, growing awareness of changing economic and social realities is evidenced by current debate concerning political devolution or independence for Scotland, and by the entry of the UK into the European Economic Community. The present

oil industry must, therefore, be viewed in this context. It has come to Scotland at a very significant time.

Development of the industry is characterised by four recognisable stages. Geophysical exploration, almost entirely an offshore activity, is followed by test drilling which requires substantial onshore service facilities. Discovery of oil and gas then requires the building of production installations, including pipelines and platforms, which provide markets for manufacturing industry. Finally, there is the extractive phase with opportunities for refining and petro-chemical industries. In practice the phases overlap considerably and before the end of 1975 all should be represented in Scotland.

The impact of the oil industry depends very largely on the rate of exploitation, and this in turn is largely decided by government policies on licensing and taxation. Whereas the effects of the Industrial Revolution were spread over several generations, the impact of the oil industry will be fully realised within one. There is, therefore, a natural concern that the environmental and social damage which accompanied nineteenth-century industrial developments should not be repeated today, despite economic and political arguments in favour of rapid exploitation. Planning procedures have particular significance in the face of such pressures and attempts have been made to modify them in order to allow schemes to proceed quickly. The problems which have arisen so far emphasise the need to control development in the interests of both the industry and the people of Scotland. In this, final responsibility obviously rests with the government of the day.

2   The World Oil Situation

During the 1960s the states of Western Europe one by one committed themselves to oil as the major source of inanimate energy. The UK was one of the last past this post when, in 1971, oil represented over 50% of its energy supply for the first time. This move away from a coal-based economy meant the increasing dependence of Western Europe on oil imported from the Middle East; a politically volatile area from which the last vestiges of European political and military influence were fast disappearing. At the time there was little reason to believe that any substantial sources of supply would be discovered in or off Western Europe.

In retrospect such a situation looks fraught with economic and political danger, yet oil did not figure as a major political issue and there was little talk of conservation of supplies. The supply and demand crisis has come ironically at the time when Western Europe, and particularly the British Isles, is on the verge of obtaining a substantial proportion of its demand from its own offshore oil and gas fields.

The aim of this chapter is to view North Sea discoveries in a world context so that their significance may be assessed in both the short and long term.

1. The Present Position

The principal elements in the world oil situation are shown in Fig. 2.1, where the total cumulative production of each major producing area is shown above the line, while below are shown the proven reserves.

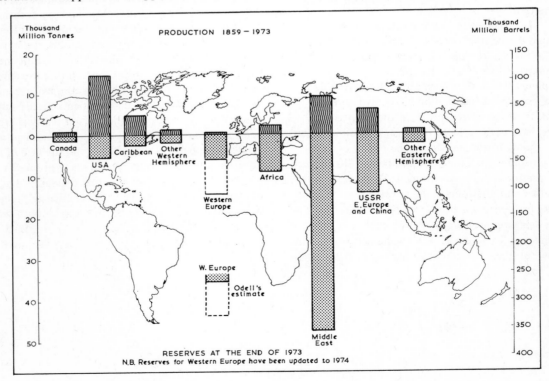

FIG. 2.1   TOTAL DISCOVERED OIL

At first proven reserves look very healthy compared with total production to date, but it must be realised that the growth in production and consumption has increased exponentially since the 1940s and thus production in 1974 was approximately eight times that of 1940. The estimated production for 1980 is fifteen times the 1940 figure. Such a growth rate cannot persist for long. The time during which quantities are doubled becomes progressively shorter, and the amounts to be doubled become astronomically large. These are subjects for further consideration.

The distribution of reserves (Fig. 2.1) indicates that the future of the world oil industry is increasingly dependent upon the Arab states of the Middle East and, to a lesser extent, North Africa. In terms of published reserves, the Middle East is followed by the Communist World. As far as the West is concerned, here lies the biggest question mark of all. It is doubtful whether Western knowledge of Siberian and Chinese reserves is accurate, and geologically these are areas where totally new fields of substantial size might be discovered.

Two of the traditional large producers, the USA, whose total production easily outstrips all others, and the Caribbean, have very limited reserves despite extensive exploration, unless tar sands and oil shales are included. Although both Alaskan and North Sea finds are vitally important new sources, coming at a particularly opportune time for the countries concerned, they do not significantly alter the balance when reserves are viewed in a global context. At best they buy invaluable time while energy policies and consumption patterns are radically altered in the Western World.

The production and consumption situation in 1973 is shown in Fig. 2.2. The Middle East had 989 million tonnes of oil available for export in 1973 while the other three major exporting areas, North and West Africa and the Caribbean, contributed 457 million tonnes. All these areas have low indigenous consumptions on account of low population totals, lack of industrialisation and, in some cases, low levels of economic development.

In contrast, Western Europe consumed 748 million tonnes and Japan, the world's second largest single importer of oil, 283 million tonnes. Japan's lack of coal reserves following the loss of Manchuria in 1945 led it to follow the same path towards dependence on oil as Western Europe, and by 1973 oil represented over 60% of its energy consumption.

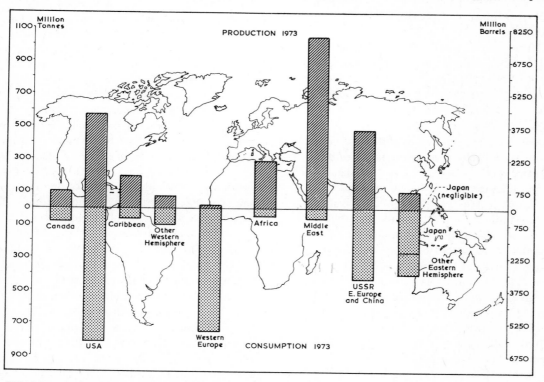

FIG. 2.2   PRODUCTION AND CONSUMPTION IN 1973

The world's largest importer and consumer, the USA, has followed a rather different path. Although a net importer since 1948, and the financial and technological power behind the opening up of the great Middle Eastern fields in the 1950s, the US reduced its dependence on imports drastically from 1959 until 1970 through an import quota system. Although not completely abandoning import quotas until 1972, the US entered the world oil market in force after 1970 and imported 313 million tonnes in 1973. This has had far-reaching consequences for the world market, and especially for Western Europe and Japan. It could also be seen as triggering off other rapid changes from the placid conditions of the 1950s and 60s to the traumatic ones of the 1970s.[1]

2. The Future

Before examining the events and circumstances that make up the current oil crisis, it is valuable to look in more detail at the question of the long-term future of the oil industry because this is intimately connected with the present situation, especially the need to conserve supplies.

The major oil companies[2] have always presented a rather pessimistic view of future supplies in relation to growing demand. For example, in late 1972 BP forecast[3] that demand could not be fully met by 1978 if it continued to increase at $5\frac{1}{2} - 6\frac{1}{2}\%$ per annum, as it had been doing. Further, after 1980 world oil production would decline absolutely as well as relatively. Mr. P. Walters, the managing director of the same company, in a recent address,[4] pointed out that oil companies feel that a reserves to production ratio of 20 years is necessary, that is that there should be at any given time proven reserves for 20 years' production at current levels. He went on to show that the annual rate of discovery from early this century has averaged 18 thousand million barrels (18×10^9) per annum. Walters regards it as optimistic to hope to maintain this rate in future, despite improved technology and offshore exploration. To make matters worse, the annual production rate for world oil now runs at 20 thousand million barrels; thus oil exploration needs to be more successful than hitherto to maintain a 20 year reserves to production ratio.

However, Fig. 2.3 shows that there are more optimistic viewpoints than those expressed above. The oil companies, and official thinking generally, estimate an ultimate total of 1 600 to 2 100 thousand million barrels (1.6 to 2.1×10^{12}) of recoverable oil in the world. It is worth noting that this ultimate figure has risen from an estimate of 500 thousand million barrels in 1940.[5] It is this rising total of both proven reserves (Fig. 2.3) and estimates of reserves ultimately recoverable with the passing of time that is the basis of a challenge to the official viewpoint from Professor P. Odell.[6] Odell postulates that the total quantity of oil that can be recovered ultimately is approaching four billion barrels (4.0×10^{12}); but that this estimate will not be established by conventional extrapolation from exploration until the 1990s. In other words, he maintains there are still vast unknown reserves to be discovered, comparable to Middle Eastern finds of the late 1930s to early 1950s.

It is not only at the global level that Odell seriously questions official estimates. Fig. 2.1 shows his upward revision of the North Sea estimates by a factor of between two and three, that is a rise from the official estimate of 44 thousand million barrels to 79–138 thousand million. [7]

How can there be such diversity in the estimates from people who have worked and studied in this field for many years? It is the more surprising when the crucial nature of the commodity is taken into account and the fact that exploration has always been characterised by the use of the most advanced technology and scientific back-up. The divergence between the oil companies' estimates and Professor Odell's is basically the difference between a geological assessment of the situation and a statistical, economic extrapolation from existing evidence.

In the 1950s it was hoped that there would be a breakthrough in the identification of the actual oil within favourable structures by seismic methods. This did not happen. Despite considerable advances in identifying structures by seismic survey, the oil geologist still depends upon drilling expensive holes to prove the existence of gas or oil. The geological view is that nearly all of the basins likely to be even remotely suitable have already been adequately explored. A major find comparable to another 'Middle East' can therefore be discounted except perhaps in remote parts of the USSR or China.[8]

The oil companies point out that not only have the most suitable geological areas been explored, but also all the easily accessible ones. The costs of

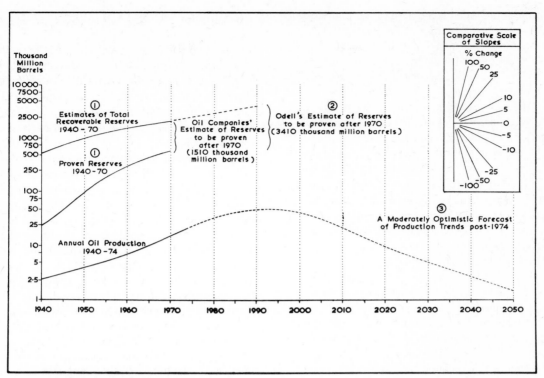

FIG. 2.3 RETROSPECT AND PROSPECT

N.B.

1. The statistics for both proven reserves and total recoverable reserves are speculative, the latter highly speculative. Thus these figures should be regarded as no more than a very general guide.
2. While the Oil Companies' estimate of a possible 1 510 thousand million barrels, to be added to the 1970 proven reserves total, is an ultimate figure in their view, Odell suggests that they will continue to revise their estimates upwards until about 1993. By that time a further 3 410 thousand million barrels will have been added to the 1970 proven reserves total. Both estimates should be treated in the light of footnote 1 above.
3. This line is conjectural and represents a median position between the ultra-pessimistic and ultra-optimistic forecasts.

developing in deep waters offshore, or in remote Arctic or tropical forest areas, are high in terms of both money and time compared to fields exploited to date. North Sea and Alaskan oil cost about ten times as much as that from the Middle East,[9] and it takes six to seven years from licensing to bring a new area into production.[10] The total development costs of known reserves over the next 10 years have been put at US $200 thousand million.[11]

The economic-statistical approach is rather different in that it virtually ignores all the geological and geophysical arguments, and concentrates instead upon past forecasts of estimates for particular fields compared with the actual reserves ultimately proved in them. This gap between the estimates, which have always proved to be on the low side compared with what was obtained, is used as the basis from which

to extrapolate and forecast the future position for new fields such as those of Abu Dhabi and the North Sea.[12] Odell states that if the world has proven reserves sufficient for 15 years' production, this is sufficient to cope with an increase in demand of 7% per annum. He maintains that the present 20 year reserve position shows that too much money has been spent in exploration and field development.

This difference of opinion between Odell and the petroleum 'establishment' about the long-term future has been considered at some length, because, according to which one believes to be the more accurate, strategies to be adopted differ markedly. The pessimistic view sees world resources of oil virtually extinguished by the year 2010—an uncomfortably close date. With oil supplying about 48% of world primary energy demand in 1974, a drastic reorganisation of life-styles is required in both developed and under-developed countries. A desperate crash programme for nuclear power and a massive revival of the coal industry would seem to be priorities for governments of all industrialised nations if the world faces declining oil production after the 1980s.

The oil companies must press ahead with the increasingly expensive deep-water exploitation of oil wherever it can be found, be it off Spitsbergen or Antarctica. The oil sands of Athabasca[13] and of other areas such as Venezuela and Colombia would then become an economic proposition, although oil produced from them would be approximately $2\frac{1}{2}$ times as expensive as that of the North Sea and 25 times that of the Middle East.[14] The oil shales of the western USA also become of interest to the oil companies and the US government, despite the massive environmental destruction involved and the fact that the energy consumed in extraction might come close to the energy obtained.[15]

If their oil reserves are to be exhausted by the end of this century the producers of the Middle East and elsewhere must immediately go for a rapid diversification of their economies. Kuwait, for instance, may have only 15 years more as an oil and gas producer. The cost of the economic transformation of these countries must be paid for by their obtaining the highest possible prices for oil over this short boom period. To buy a little more time these producers might be well advised to adopt conservation measures by artificially restricting production.

A slightly less pessimistic view is also shown on Fig. 2.3 in which a production peak is reached in the 1990s and the present level of production of 16 to 20 thousand million barrels (16 to 20×10^9) is reached again in about 2010. The most optimistic view, based on the assumption of total reserves of 4 000 thousand million barrels, however, postpones the need to replace oil as a major fuel until the second quarter of the 21st century and oil remains significant in the world economy until about 2050. This latter view incorporates the premise that the demand for oil does not increase exponentially indefinitely, but rather that demand is eventually satisfied at a high level of consumption. After that the rate of increase falls to about 3% per annum instead of 6%–7% as at present. Obviously if one subscribes whole-heartedly to the optimistic view the present crisis is a temporary one and there is no need for desperate crash programmes in coal, nuclear power or tar sands and oil shales.

3. The Oil Crisis

Five years ago few people had heard of OPEC, although it had been formed in 1960. Since October 1973 the Organisation of Petroleum Exporting Countries has featured prominently in the news. The rest of the world awaits its meetings with anxiety, or simply resigns itself to the seemingly inevitable rounds of price increases which result.

The organisation was formed so that the producers, a group of weak divided states, could achieve through concerted action a more favourable deal from the giant oil companies. In the late 1950s and early 1960s there was a world surplus of oil and the international oil companies could keep prices low by playing one producer off against another. The price of oil was prevented from rising above about US $1.50 per barrel, which in real terms meant it was falling in price. After a period of sharply rising oil imports the USA had imposed a quota system in 1959 which largely closed that market to foreign producers and helped depress world oil prices throughout the 1960s. It was under these favourable conditions that Western Europe ran down its coal industry and swung over to oil and to dependence upon the Middle East.

By the end of the 1960s rising world demand for oil (especially from Western Europe and Japan), together with the growing inadequacy of US oil production in the face of expanding domestic consumption, drastically reduced world oil surpluses and changed a buyers' into a sellers' market. In 1970 OPEC managed for the first time to achieve unity of purpose among its members who began progressively to raise the 'posted' prices for crude oil and thus their revenues.[16]

Libya, by imposing restrictions on production in 1971, was the first OPEC state to challenge successfully the supremacy of the international oil companies. It could particularly afford to do this because of its attractiveness to them both in terms of its location (with the Suez Canal closed) and the high quality of its light crudes. This was followed in 1973 by the expropriation of the US company's (Occidental's) assets by the Libyan Government. In the same year the US reopened its market to unrestricted oil imports, still further strengthening the control which OPEC was beginning to exert over the world oil industry.

Since then producers have insisted on increasing control over the rates at which their resources are being exploited. The formation of national oil companies has led to control of oil concessions in individual countries of up to 100%.[17] The international companies develop the fields, but the host governments receive a substantial proportion of the production through their national company. This oil may be disposed of as the host government sees fit. Sometimes it is auctioned on the world market.

In October 1973, the crisis finally broke as a result of the renewed Arab–Israeli conflict and the action of Arab states in drastically reducing oil supplies to the US and other major importers alleged to be helping Israel. By November 1973, production from the Arab states was 25% below the September figure. These cuts were rescinded in the spring of 1974 but the threat of their reimposition remains.

Between October and December 1973 the posted price of oil from the Persian Gulf quadrupled and reached nearly US $12 per barrel. Even spread over five years or so such rises would have been serious to a world committed to economic development based on cheap oil. Imposed in three months they were catastrophic. In the dramatic words of Peter Hillmore, 'the 12 horsemen of the OPECalypse rode into general view, and the world as we knew it ended.'[18]

The economies of Western Europe and Japan have had crippling burdens of oil payments placed upon them.[19] This problem has inevitably preoccupied them since October 1973 because serious financial collapse, on account of high prices, among the richer nations could drag the whole world economy into the abyss.

The effect of rocketing oil prices on the under-developed world has also been serious. High prices for energy and petro-chemicals may wipe out all the fragile economic achievements of the last twenty years and reduce the already slim chances of avoiding severe famine in many areas. Expensive oil could absorb all their foreign exchange reserves, merely to maintain their present levels of imports. Economies would stagnate or regress if nothing was done.

The situation at the beginning of 1975 is highly unstable. Oil prices are still rising despite the expressed wish of Saudi-Arabia, the largest producer, to hold them or even bring them down. The income pouring into the coffers of oil producers has reached an almost meaningless level. In 1974 their investment surpluses amounted to US $60 thousand million and, without further price increases, could read US $600 thousand million by 1980. Such funds are almost beyond the range that world financial institutions can handle without provoking runaway inflation. Countries with strong currencies do not want their economies threatened by these vast sums. The US government is trying to force down oil prices

and if this money were poured into the USA it could possibly be used as a weapon by the US government against the producers. In consequence, one of the few remaining outlets is London, surprising in view of the currently weak economy of the UK. However, the effects of such large-scale investment could be to mortgage the UK economy to the Arab world.

In effect the high-cost oil situation has created a vicious circle in world economics. Developed and under-developed countries alike must buy oil at prices they cannot afford. The producers must sell oil (in many cases it is all they have got), but they end up with concentrations of money greater than the world economy can effectively absorb. There is also the added danger that should the situation become more desperate, Western countries might be tempted to expropriate oil producers' investments within their respective territories.

Viewing this conflict between OPEC and its customers as interested observers, and participating politically and to a limited extent as suppliers, are the USSR and to a lesser extent the rest of the Communist World. The USSR, with adequate oil reserves plus 60% of the world's coal reserves, is politically and economically in a very strong long-term position. Almost alone its policies and economy over the next decade do not need to be dictated by the need to ensure energy imports and overcome financial problems associated with them.

For the world at large the situation seems grave for the next few years, but after that changes in energy policies and new discoveries could ease the situation. There seems to be general agreement, in theory at least, that there must be properly concerted policies for energy conservation. In practice agreement on such policies may be hard to reach, even attempts to present a united front by consumers towards OPEC. At present (January 1975) the world's largest consumer, the USA, which could make the biggest contribution to energy conservation and thus to the easing of pressure on supplies, is still basically pursuing a cheap energy policy within its huge domestic market. The Chairman of BP, Sir Eric Drake, sees a need not only for the measures outlined above, but also for full consultation between OPEC, the oil companies and the consumers to reconcile points of view.[20]

The chapters which follow consider in detail oil and gas developments in the Scottish area of the North Sea, but how significant are they in terms of this world situation? Even the most optimistic estimate of North Sea reserves does not greatly alter the overall world picture, but in the shorter term the North Sea can be critical for Western Europe. The reserves of the North Sea and Alaska taken together give the West a valuable bargaining counter in negotiations with OPEC. The resources of these two oil provinces may help in reintroducing financial stability to the world oil market.

It is difficult to envisage a world without significant quantities of mineral oil, yet the least optimistic forecasters suggest this will happen before the present school generation reaches middle age. Even the most optimistic assessment is that oil supplies will run out by the middle of the 21st century. Whichever proposition is correct, it is fully time that governments throughout the world gave higher priority to preparations to meet this contingency.

REFERENCES AND NOTES

1. For an account of the background to the post-war world oil situation, see ODELL, P. R. (1974a), *Oil and World Power,* 3rd Edition, Pelican, London.
2. The seven 'Majors' are: (i) Standard Oil of New Jersey (Exxon); (ii) Royal Dutch Shell; (iii) British Petroleum; (iv) Standard Oil of New York (Mobil); (v) Standard Oil of California (Chevron); (vi) Gulf Oil; (vii) Texaco. These companies dominate the non-Communist oil world as producers (80%), refiners (70%) and tanker fleet operators (50%). ODELL, P. R., *op. cit.,* pp. 11–12.
3. WARMAN, H. R. (1972), 'The Future of Oil,' *Geogr. J.,* 138 (3), pp. 287–297.
4. WALTERS, P. (1974), 'Carbon and hydrogen sources – the Supplier,' reprinted from *Pet. Rev.,* June 1974.
5. KIRKBY, M. and ADAMS, T. (1974), *The Future for Oil in 1999,* notes issued by British Petroleum. See also ODELL, P. R. (1974b), *Energy Needs and Resources* (Aspects of Geography Series), p. 25, Macmillan, London.
6. ODELL, P. R. (1974b), *op. cit.*
7. ODELL, P. R. and ROSING, K. (1974), 'Weighing up the North Sea Wealth,' *Geogr. Mag.,* 47(3), December 1974, pp. 150–5.
8. KIRKBY, M. and ADAMS, T. (1974), *op. cit.;* WHITEMAN, A. (1974), Inaugural lecture, Aberdeen University, December 1974; personal conversations with oil geologists and geophysicists in Aberdeen.
9. WALTERS, P. (1974), *op. cit.*
10. *Ibid.*
11. *Ibid.*
12. ODELL, P. R. and ROSING, K. (1974), *op. cit.;* ODELL, P. R. (1974b), *op. cit.,* pp. 21–22.
13. ANON. (1973), 'Oil sands of Athabasca,' *Shell Education News,* Summer Term, 1973.
14. WALTERS, P. (1974), *op. cit.*
15. WINCHESTER, S. (1974), 'Shale goes stale,' *The Guardian,* 12 November 1974.
16. The posted price of oil is the price set on oil crossing the border of a producing country. Until 1970 this price was fixed by the oil companies, from 1970–73 by OPEC and the oil companies, and since October 1973 by OPEC alone.
17. Each OPEC member now holds at least a 60% stake in oil company operations in its territory, in addition to control of concession areas.
18. HILLMORE, P. (1974), 'Year of the Sheikh,' *The Guardian,* 17 October 1974.
19. AUS$1 per barrel increase in oil prices adds £400 million to the UK balance of payments deficit.
20. DRAKE, E. (1974), Address to the annual meeting of the American Petroleum Institute, 11 November 1974.

3 The Physical Setting

The interaction between the oil industry and the environment is dominated by three main themes. These are, the search for petroleum concerned mainly with elucidating the subsea geology, problems posed by the marine environment for offshore engineering and geographical patterns arising from the physical nature of neighbouring coastlines. Initially, the search for oil and gas is directed largely towards ascertaining the geology beneath the sea bed. Although this work continues throughout the development of an offshore petroleum province, the early stages of the production phase now in progress are more preoccupied with the marine environmental problems of offshore engineering. A whole new technology for the extraction of oil and gas in waters deeper than ever before exploited anywhere in the world is currently being developed off North West Europe. The aspect of this offshore work which impinges most directly on ordinary people is its impact on coasts. Coastal communities may suddenly find themselves confronted by schemes for pipeline landfalls and terminals, service bases, platform building yards and, perhaps, even refineries.

1. The Great Search: Geology, Exploration and Fields

A. Sea-bed topography and geological structure

Topographically, the sea areas around Scotland are divisible into three parts—the continental shelf above 200 m, the intermediate depths (notably the continental slope) between 200 m and 2 000 m, and the deeps below 2 000 m. Scotland occupies a central position in relation to the continental shelf off North West Europe and the complex of intermediate depths west of this shelf (Fig. 3.1). Geologically, the continental shelf is an extension of the land mass of Europe. The area of the shelf proper comprises the shallow seas east of the North East Atlantic Basin which are separated from other shallow areas further west by the Rockall Trough and the Faroe—Shetland Channel.

This area consists in the main of extensive deposits of sedimentary rocks of continental and shallow marine facies, parts of which have been proved to contain very large reserves of oil and gas.

The intermediate depths consist of three basic features—the continental slope of the shelf proper, the Faroe Rise and the Iceland—Faroe Ridge. Of these, the continental slope is potentially most important in further exploration, at least to begin with. From exploration of the slope, largely undertaken for academic rather than oil-prospecting purposes, it would appear that the rocks on the eastern slopes of the Faroe—Shetland Channel and Rockall Trough are seaward continuations of the formations presently being explored for oil at shallower depths.[1] These are comparable to similar formations at equivalent depths off Norway, between the northern end of the Norwegian Deep and the Barents Sea, where at about 66°N a scientific expedition recently struck oil.[2] At present concession areas extend on to the continental slope in Scottish waters, both to the north and west (Fig. 4.5), and it appears that these intermediate depths will assume greater significance in future rounds of licensing. Although with present technology any oilfields in areas below 200 m remain largely out of reach in terms of recovery, serious consideration of these is merited in the longer term. Such areas are situated in geological formations as extensive as those presently being explored on the continental shelf. Prospects for their exploitation will increase with the further development of subsea completions and floating platforms.

The second important feature at intermediate depths is the Faroe Rise, a complex relatively shallow region extending north-eastwards from the Rockall Plateau to the banks west of the Faroe Islands, including those upon which Faroe is situated. A substantial part of this area is somewhat below 1 000 m, and is linked to the continental shelf by the Wyville-Thomson Ridge. It has now been confirmed beyond doubt that the Faroe Rise is a fragment of continental crust,[3] possessing a similar structure to the shelf proper.

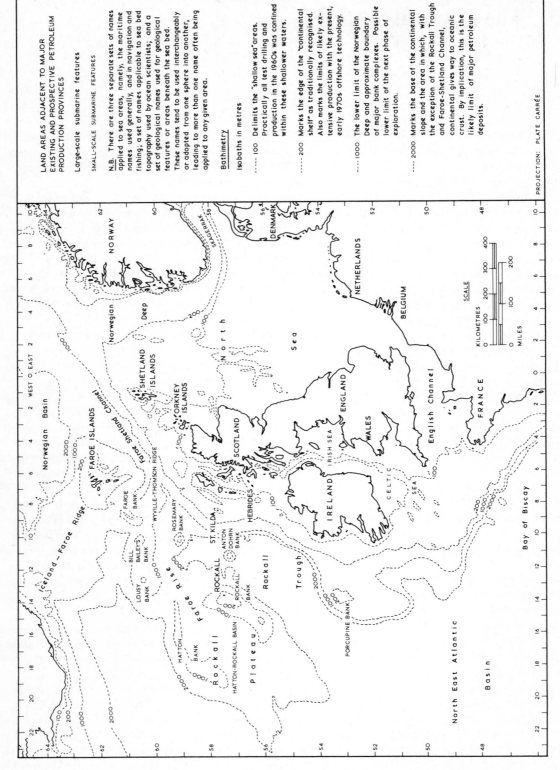

LAND AREAS ADJACENT TO MAJOR
EXISTING AND PROSPECTIVE PETROLEUM
PRODUCTION PROVINCES

Large-scale submarine features

SMALL-SCALE SUBMARINE FEATURES

N.B. There are three separate sets of names
applied to sea areas, namely, the maritime
names used generally, and in navigation and
fishing; a set of names applicable to sea bed
topography used by ocean scientists; and a
set of geological names used for geological
features or areas beneath the sea bed.
These names tend to be used interchangeably
or adopted from one sphere into another,
leading to more than one name often being
applied to any given area.

Bathymetry

Isobaths in metres

---- 100 Delimits the "shallow sea" areas.
Practically all test drilling and
production in the 1960s was confined
within these shallower waters.

---- 200 Marks the edge of the "continental
shelf" as traditionally recognised.
Also marks the limits of likely extensive production with the present,
early 1970s offshore technology.

---- 1000 The lower limit of the Norwegian
Deep and approximate boundary
of major bank complexes. Possible
lower limit of the next phase of
exploration.

---- 2000 Marks the base of the continental
slope and the area in which, with
the exception of the Rockall Trough
and Faroe-Shetland Channel,
continental gives way to oceanic
crust. By implication, this is the
likely limit of major petroleum
deposits.

PROJECTION: PLATE CARRÉE

FIG. 3.1   THE OFFSHORE TOPOGRAPHY ADJACENT TO THE BRITISH ISLES

10

This view is confirmed by evidence of the presence of deep sedimentary rock basins, notably the Hatton-Rockall Basin. Meanwhile, the shallowest areas of this bank complex comparable in depth to the shelf appear to be composed, in part at least, of igneous rocks, as are the Faroe Islands. It is generally thought that the fragment came into existence during the opening phases of the separation of Greenland, the Faroe Rise and North West Europe in Cretaceous and early Tertiary time,[4] while the extensive igneous rock outcrops belong to the broadly contemporaneous phase of vulcanism associated with the breaking apart of the continents, evidenced on land by the Antrim Plateau, parts of the Inner Hebrides, the Faroe Islands and the older rocks of Iceland.

In contrast to the two previous areas of intermediate depth, the Iceland–Faroe Ridge is an oceanic structure composed mainly of igneous rocks[5] falling within the geological province of Iceland. Despite comparable depths, it seems unlikely that there are any sedimentary basins on as large a scale as in the previous two areas and, by implication, few if any significant oil deposits are likely to exist here. It is worth noting, however, that the presence of oil was recently recorded on the Jan Mayen Ridge to the north of Iceland,[6] indicating the presence of further continental fragments in this general vicinity.

Below 2 000 m the oceanic basins of the North East Atlantic and Norwegian Sea are of little concern in the present context, as at this depth the sediments of the continental slope give way to relatively thin deposits of oceanic type, underlain by oceanic crust. The only major exceptions to this pattern exist in the Rockall Trough and Faroe–Shetland Channel, which are characterised by thick layers of sedimentary rocks, probably largely of shallow marine and continental provenance.[7]

The importance to Scotland of these extensive intermediate depths and deeps between the continental shelf and the Faroe Rise lies in their relative accessibility to Scotland. Although such areas will lie partly in Irish, Faroese or Norwegian waters (jurisdiction areas have still to be defined), Scotland may well prove to be the most convenient landfall from which to conduct both exploration and the development of any subsequent discoveries. Just as Frigg gas is to be piped to Scotland, and Ekofisk oil and gas to North East England and West Germany respectively (Fig. 4.6), so the scale of the oil industry in Scotland seems destined to be more than a direct function of developments in Scottish waters alone.

B. Exploration

Although the broad picture of sea-bed topography and marine geology outlined above had been taking shape for some time previously, the search for petroleum has greatly added to detailed knowledge in a relatively short time.[8] In the context of exploration, discussion of petroleum geology[9] is limited to a brief review of important factors governing the natural occurrence of petroleum. This is followed by a general outline of important geophysical techniques used in building up a coherent picture of geological conditions.

Naturally occurring hydrocarbons exist in two basic forms, namely liquid petroleum and natural gas (see glossary). These may occur virtually exclusively in one form or the other although there is commonly some admixture of the two. Occasionally, as for example in the Cod Field in the Norwegian sector, gas under very high pressure mixed with light-coloured petroleum is encountered. This form is known as 'gas condensate' or simply 'condensate'. In consequence, detailed specifications of discoveries usually include a reference to the gas/oil ratio (GOR), which is a measure of the admixture of crude oil and natural gas present.

It is now generally acknowledged that naturally occurring hydrocarbons are derived from the remains of living organisms having a predominantly marine or fresh water provenance, which have become concentrated in the processes of sedimentary basin formation. Thus crude oil and natural gas are normally found only in sedimentary rocks, usually where these are disposed in some kind of extensive basin structure. From the point of view of the petroleum geologist there are three kinds of sedimentary rocks of primary interest, namely source, reservoir and cap rocks. The source rocks are those in which the petroleum originates. In order for migration to occur, rocks intervening between source and reservoir rocks, where the deposits accumulate, must be permeable. While it is doubtful if any one class of rock is dominant as a source rock, clays and shales are certainly common in this context, while the underlying coal measures of Carboniferous age are the source rocks for much of the natural gas in the southern North Sea.[10] Reservoir rocks are often chalk or sandstone. The final requirement for an oil or gas accumulation is a suitable impermeable cap rock, geometrically disposed in such a way that the petroleum cannot escape. Typical cap rocks include shale over sandstone, and evaporite over limestone. Some of the most common stratigraphical conditions are illustrated in simplified form in Fig. 3.2.

Since field mapping and ground survey are impossible, preliminary exploration, that is the pre-test-drilling stage, relies more or less completely upon geophysical techniques.[11] There are three major techniques involved. The first is the airborne magneto-meter survey, which measures the magnetic properties of rocks. It is especially useful for detecting the broad configuration of the basement rocks on which sedimentary basins are founded (Figs. 3.2, 3.4), as such rocks commonly have a relatively high proportion of minerals containing iron. Magnetometer survey is a first necessity, as it makes possible separation of the major subsea areas of sedimentary rocks from similar areas in which sedimentary cover is shallow or absent. The result is an aeromagnetic map from which a rough outline of the extent of the sedimentary basins (as in Fig. 3.5) can be plotted.

The aeromagnetic technique, however, often tells little regarding the sedimentary structures as such. For this it is necessary to resort to two further techniques, gravimetric and seismic, both of which are usually carried out by survey ship. The gravimeter is used to detect sedimentary structures by means of measuring relatively minute anomalies of the earth's gravitational field produced by different types of rock. It is particularly useful in the detection of salt structures, which characteristically occur in conjunction with gravity 'lows' and are also commonly associated with oil-bearing formations due to the propensity of salt to flow under pressure of overlying rocks, and thereby to assist in the formation of suitable traps. (Fig. 3.2)

The seismic technique is by far the most precise of the three methods, and is essential for mapping sedimentary structures in detail preparatory to the selection of exploratory drilling sites. It involves setting off series of shock waves generated by small explosive charges, or otherwise creating an impact, for example by electrical means. These miniature earthquake waves are reflected and refracted as they penetrate successive strata (Fig. 3.3). Thus a relatively detailed geological picture of the stratigraphical succession is obtained at any given point. Improvements in seismic techniques effected in the course of North Sea exploration are now being utilised world-wide in offshore operations. One great advantage of using a ship for these surveys is the rapid rate at which it is possible to complete coverage

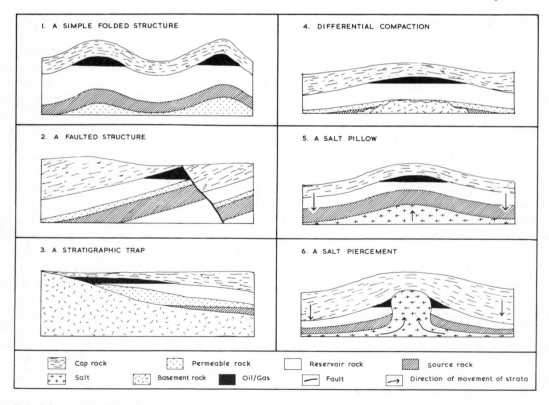

FIG. 3.2   THE GEOLOGY OF OIL AND GAS RESERVOIRS

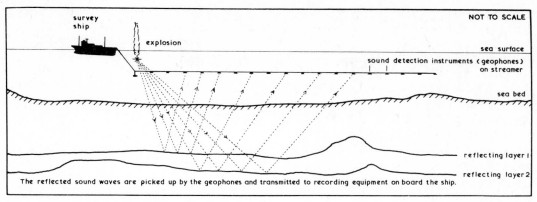

FIG. 3.3   THE SEISMIC REFLECTION TECHNIQUE

of extensive areas, in comparison with land-based surveys. Ultimately, however, there is no substitute for test drilling. At this point the reverse situation obtains as conditions at sea present greater difficulties than those on land.

C. The geology of the oil and gas fields

The end result of exploration to date has been the marking out of the major basins illustrated in Figs. 3.4 and 3.5. Broadly speaking, there are two vast areas of sedimentation to the east and west of Scotland which come together north of Shetland. Each area forms part of a much larger sedimentary province. To the east lies the North Sea province, and to the west part of an 'Atlantic' province, the other half of which lies off the coasts of East Greenland and the Canadian Maritime Provinces, separated by the phase of continental drift which began in the Cretaceous and Tertiary and which appears still to be in progress.[12] Indeed, oil prospecting has already begun off the east coast of Canada.

Within these very large basins lie a series of subsidiary basins.[13] The whole of the North Sea area, for example, can be visualised as a single basin which has been subsiding more or less continuously since Upper Palaeozoic times (Fig. 3.4), different areas subsiding at variable rates in successive periods to produce these secondary basins. It is in these second-order basins, and in structural features within them, that exploration interest is centred. Indeed, in the absence of detailed public knowledge of exploration findings, the best way in which the uninitiated can assess the possibilities of discovering oil is to note the pattern of licence distribution in successive rounds of allocations.

These reflect reasonably well the pattern of major prospective basins (Figs. 3.5, 4.5).

In the area of the North Sea within the UK sector there were during the Permian and Triassic periods two major areas of subsidence in the north and south respectively, separated by the Northumbrian Arch and the Dogger Bank High. In the southern basin of Permo-Triassic age lies the first of the three major areas of petroleum deposits discovered to date, namely the gas fields of the southern North Sea.[14] Discovered mainly in the late 1960s, the gas is derived from Carboniferous source rocks, and held chiefly in sandstone reservoirs belonging to the Rotliegendes formation of Permian age, capped by Zechstein salt deposits, the movement of which under pressure of overlying rocks has produced many structures suitable for the entrapment of gas (Figs. 3.2, 3.4, and 3.5).

This basin configuration was superseded by a single large subsiding feature in Late Cretaceous and Tertiary times, extending more or less down to the present. In this thick sedimentary sequence of predominantly shallow marine facies have accumulated most of the oil reservoirs characteristic of the remaining two major petroleum basins so far discovered in the North Sea. The more southerly of these, generally known as the Central North Sea Basin, is faulted producing a graben-type structure. The middle of this basin coincides roughly with the median line between Scottish and Norwegian waters. The fields in the southern part of the basin include the Ekofisk complex and the Dan field, where the main reservoirs are Upper Cretaceous Danian chalk formations, fractured by upwelling of Zechstein salt from below. The Auk and Argyll fields lie at older Permian levels, while further north the Forties and Montrose reservoirs are sandstones of Palaeocene (early Tertiary) age. The

13

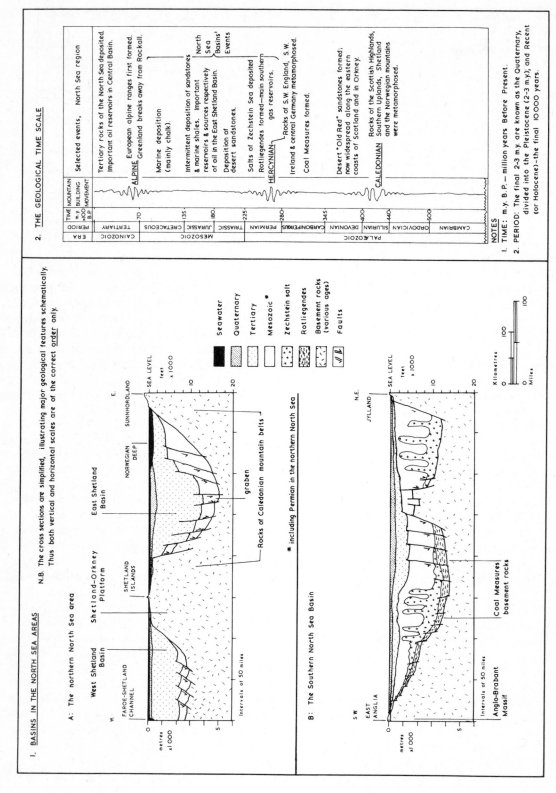

FIG. 3.4 THE STRATIGRAPHY OF THE NORTH SEA

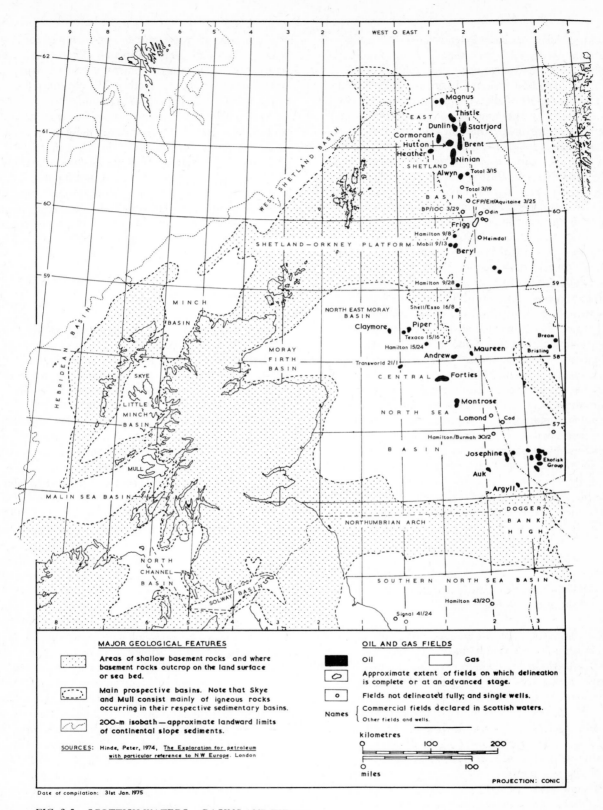

FIG. 3.5   SCOTTISH WATERS – BASINS AND FIELDS

Central North Sea Basin extends north-westwards into the Moray Firth area, where the Piper and Claymore fields are situated in Jurassic sands.[15]

The East Shetland Basin, currently proving to be the richest of all, is characterised mainly by oil fields together with a number of gas accumulations at several horizons.[16] While Jurassic shales are probably the principal source rock, the most common reservoirs in this thick marine sequence are Jurassic sands as in the cases of Brent, Hutton and Ninian, or Palaeocene sandstones as in Frigg. Some of the oilfields discovered to date contain gas in commercial quantities as well, notably the giant Brent Field (Fig. 4.6).

To the west of Scotland the second large province appears to contain a more complex series of basins, as yet little explored. These occur to the west of Shetland and Orkney, in the Minches and in the areas between the coasts of Scotland and Ireland.[17] Traces of oil have been found in test drilling west of Shetland,[18] and there are high hopes of finding petroleum both here and on the slope beyond. The Scottish—Irish slope consists of a complex series of shelf-type deposits produced by erosion of adjacent land masses, the rocks being comparable in age to those in the North Sea basins. The beds are generally seaward-dipping, sometimes convoluted as they 'spill' over one another down the relatively steep continental slope. The instability of the basement rocks of this slope, characteristic of such relatively young continental margins, has resulted in extensive faulting accompanied by the creation of basement ridges and basin-like structures. Many of these are potentially oil and gas bearing.[19]

2. The Maritime Frontier: Marine Environment and Offshore Engineering

The marine environment interacts with and influences the technology of the oil industry in all phases of offshore development—exploration, construction of harbours and service bases, operation of drilling rigs, installation of pipelines and platforms—and during the production phase (Fig. 5.7). Among the environmental factors requiring consideration are atmospheric and sea surface conditions, notably winds and wave patterns, but also including currents, tides, surges, air and sea temperature conditions and the possibilities of encountering snow and ice. In the solution of many practical engineering problems, wave and current

conditions in particular must be considered in conjunction with water depth, also an important factor in its own right. It is thus convenient to discuss winds, waves and water depths together. The characteristics of the sea bed zone are also of great importance, especially soil conditions and bottom currents. It is perhaps most instructive to consider these environmental factors in the context of three broad areas of engineering operations.[20] These are the exploration stage, especially exploration drilling and supply vessel operation; methods of oil and gas extraction, including production platforms and subsea installations; and transportation problems, notably tanker loading in the open sea and the laying of pipelines.

A. Winds, Waves and Water Depths

The rapid variability of, and extremes in, weather conditions[21] were perhaps the first features to impress themselves upon the offshore industry. In the late 1960s, no less than three rigs (*Sea Gem, Constellation* and *Ocean Prince*) were lost while drilling for gas in the southern North Sea; while in the winter of 1973—74 *Transocean 3,* a brand-new semi-submersible, was lost in the vicinity of the Piper Field. Added to these many lesser mishaps and interruptions due to bad weather have accumulated into large quantities of 'downtime' during which drilling is suspended, extending to between 10 and 20% of the time taken to drill a well.[22] With rig operating costs in the range of £10 000 to £20 000 per day, such delay means vast expense. Bad weather also disrupts shipping links, especially the working of supply vessels close to rigs and entry into exposed east coast port approaches. Shipping commonly traverses several 'regions' of combined weather and sea conditions on passage to and from rigs.

Although broad weather patterns on a North Atlantic scale can be predicted with reasonable accuracy, local configuration of land and sea, and local meteorological and hydrographical peculiarities are of first importance in understanding and coping with environmental conditions encountered in specific areas. Indeed, regional geography is the key not only to weather forecasting,[23] but also to the totality of environmental factors encountered in offshore operations. For example, in the southern North Sea and areas of comparable depth or close to coasts, wave heights are often accentuated by relatively shallow water and strong currents. In the northern North Sea and to the west of Scotland, swells originating in Atlantic storms far removed from local conditions are common, and higher and longer waves are produced (Fig. 3.6). These swells are frequently complicated by local wave trains generated by fast-moving depressions and other relatively localised synoptic features, leading to complex rapidly changing wave patterns. Wave data for the North Sea and Atlantic around Scotland are very limited.[24] In order to remedy this a £600 000 research scheme is being sponsored by the oil companies and the Department of Energy to collect meteorological and hydrographic data on a

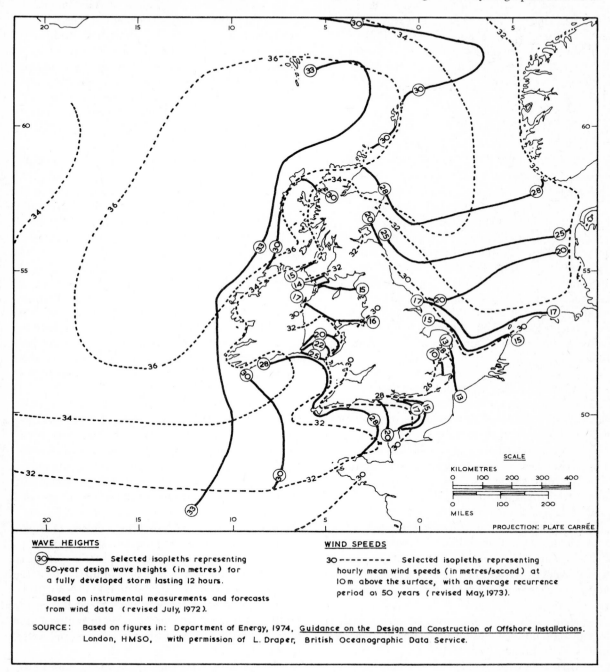

WAVE HEIGHTS

— Selected isopleths representing 50-year design wave heights (in metres) for a fully developed storm lasting 12 hours.

Based on instrumental measurements and forecasts from wind data (revised July, 1972).

WIND SPEEDS

----- Selected isopleths representing hourly mean wind speeds (in metres/second) at 10 m above the surface, with an average recurrence period of 50 years (revised May, 1973).

SOURCE: Based on figures in: Department of Energy, 1974, Guidance on the Design and Construction of Offshore Installations. London, HMSO, with permission of L. Draper, British Oceanographic Data Service.

FIG. 3.6   OFFSHORE WIND AND WAVE PATTERNS

long-term basis. Two of the three research vessels involved are stationed in Scottish waters near Shetland.

Another vital marine factor is the depth of water (Fig. 3.1). In the early days of North Sea exploration when activity was concentrated in the southern waters this did not present insurmountable difficulties. Operation in comparable depths had been carried on for a long time, notably in the Gulf of Mexico. Water depths were generally less than 90 m (300 ft), and the use of jack-up rigs was the rule. North of 57° N, however, the sea deepens to the 90—180 m range (300—600 ft). This has necessitated the use of semi-submersible rigs designed to drill in depths of 180 m and more. The expected continuation of drilling northwards and westwards beyond the shelf edge, both in the North Sea and elsewhere, has led to the development of a new generation of semi-submersible rigs capable of operating in water depths of 300 m (1 000 ft), while many rigs are being modified to drill in 900 m (3 000 ft) depths. It is considered that, despite mooring problems in such depths, these rigs will still be competitive with the new generation of large drill ships,[25] twenty-five of which have computer-controlled thrust systems for maintenance of position during drilling, and can operate in depths of around 900 m or more. These developments have brought the upper half of the continental slope well within range of exploration drilling.

Meanwhile, in the context of production installations, the greater water depths and severe weather conditions experienced in the northern North Sea have led to the design of novel types of production platform.[26] The currently building steel and concrete types are in principle scaled-up versions of earlier designs. They depend like the latter entirely on the sea bed for their support (see below) either in the case of steel structures by piling, or in the case of concrete designs by relying simply on gravity (Fig. 3.7). Such platforms are suitable for water depths up to the 180 m (600 ft) range. In deeper water problems of stability, massive materials requirements and consequent high costs favour designs which are inherently buoyant and are anchored to the sea bed. The leading examples are designs based on the tension-leg principle, in which anchoring cables are maintained in a taut state through tension applied by a submerged buoyant platform base (Fig. 3.7).[27]

To cope with wind and wave conditions, rigs and platforms are currently being designed to withstand wind speeds of the order of 67—77 m/s (130—150 knots) and wave heights in the 30 m (100 ft) range.[28] Although detailed design criteria can be expected to vary quantitatively for any specific location (note variations in Fig. 3.7), the wind, wave and current loadings on platforms are both complex and substantial. With this in mind, certain minimum conditions have been recommended for special attention in assessment of environmental conditions for design purposes.[29] These include the maximum wave height likely to be recorded in a 50-year period ('the 50-year design wave') (Fig. 3.6); and the maximum hourly mean wind speed likely to be encountered in a period of not less than fifty years, an important consideration in the calculation of the maximum design wave height.

The '100-year storm' conditions with which most designs are intended to cope[30] are far more severe than originally envisaged or encountered to date in any other offshore drilling region. Further, duration of high winds and seas for periods in the order of 25% of the time in northern waters in winter are a cause for concern. Prolonged exposure to waves smaller than storm waves but more frequent in occurrence may be more damaging due to the greater effectiveness of these waves in promoting fatigue in offshore structures. In such extreme conditions, it becomes imperative to 'work with nature' as much as possible. Apart from tension-leg designs, novel ideas intended to minimise the effects of wave action on stability include the 'tuned sphere drill ball' concept, considered to be applicable both to rigs and platforms.[31]

In order to avoid surface problems associated with wind and wave action, and the navigational hazards which platforms present, an alternative is the use of subsea production systems which enclose well heads and other production equipment in modules placed on the sea bed. This reduces the problems, but the continued necessity for a surface support ship does not necessarily eliminate bad weather interruptions. Moreover, these systems avoid the problems of large-scale structural engineering required for platform building. Although some sixty or so subsea systems are in use around the world, in depths of up to 110 m (360 ft), and this type of well completion has been in use in the Ekofisk complex for three years,[32] much design work remains to be done to reduce further operational problems associated with sea surface conditions and water depths.

A further difficulty in the use of subsea production systems is that it is possible to sink only a small

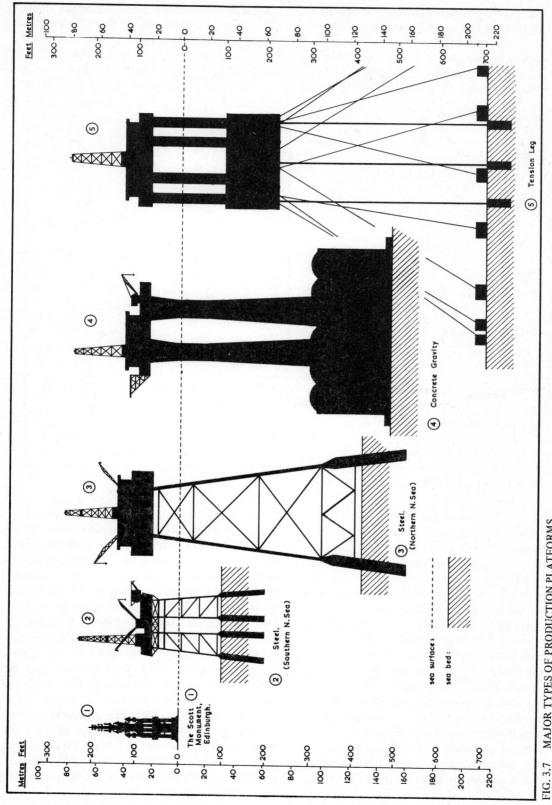

FIG. 3.7 MAJOR TYPES OF PRODUCTION PLATFORMS

number of wells—one to five—from designs currently in use, while over forty deviated wells (that is wells drilled at angles other than the vertical) can be sunk from a major production platform, drawing petroleum from a large area in order to achieve maximum recovery of reserves (Fig. 3.8). In the Forties Field,

feasible method. One of the attractions of concrete platforms is their individual built-in storage capacity for up to a million barrels of oil. To date, separate large-scale offshore storage facilities have been adopted only on the Ekofisk Field. Large storage units are necessary to match production to removal by tanker

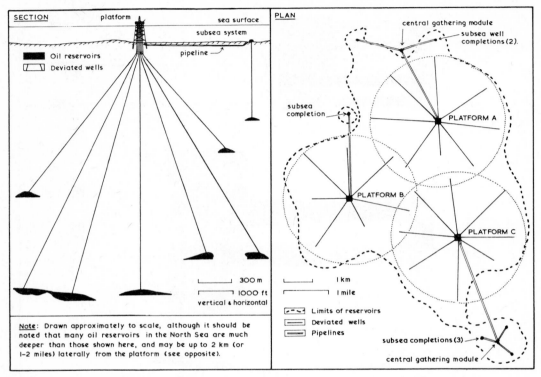

FIG. 3.8   EXPLOITATION OF AN OILFIELD

for example, twenty-seven deviated wells from each of the four platforms will terminate at maximum lateral distances in the order of $2\frac{1}{2}$ km ($1\frac{1}{2}$ ml) from the platform.[33] Also, apart from the fact that difficulties are likely to be encountered in operating it at great depth, the complex equipment necessary for large scale production is too large for accommodation in existing subsea systems and must therefore be housed on the platform.[34] This applies notably to the plant for separating gas and water from crude oil.

The close connection between surface conditions and water depth in practical situations is further illustrated by the problems posed in transporting oil from the fields to the shore. Basically there are two alternatives—either to load the oil into tankers at sea, which usually requires some kind of offshore storage facility, or to pipe it ashore. In the case of gas in large quantities, pipelines are the only

or pipeline, but these are tied to one specific location and are uneconomic in the long run for small fields.

For such fields, and sometimes also in the early stages of production from major fields, some kind of floating storage system or direct tanker loading is preferred. A leading example of the former is the SPAR to be used in the initial stages of production from Brent. With a capacity of 300 000 barrels, it has the advantage of mobility over platform storage, allowing it to be moved elsewhere when no longer required at Brent.[35] The single buoy mooring system for loading tankers direct from wells has already been used fairly widely around the world: the Ekofisk extraction system includes examples but it has been shown that the severe conditions experienced in the North Sea require the use of scaled-up versions, such as that placed on the Auk Field (Fig. 3.9), often referred to as Exposed Location (ELSBM) types. Like

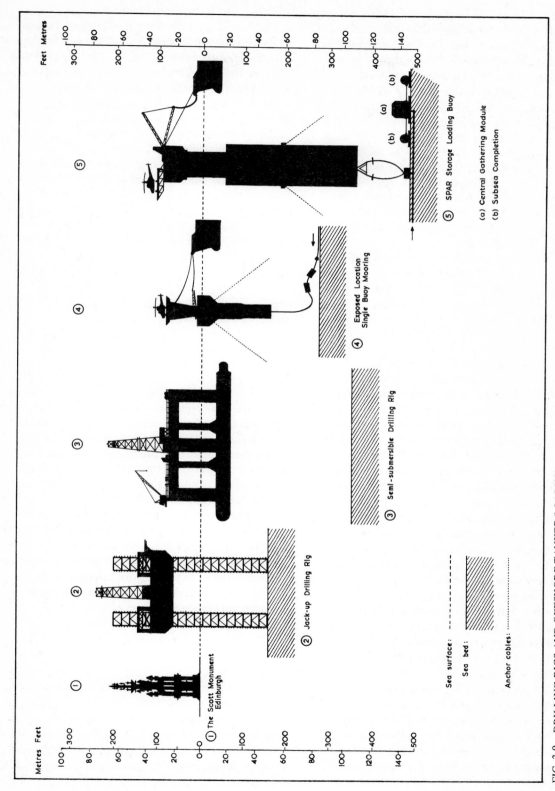

FIG. 3.9   DRILLING RIGS AND OFFSHORE TANKER LOADING

① The Scott Monument Edinburgh

② Jack-up Drilling Rig

③ Semi-submersible Drilling Rig

④ Exposed Location Single Buoy Mooring

⑤ SPAR Storage Loading Buoy

(a) Central Gathering Module
(b) Subsea Completion

Sea surface: _____
Sea bed: ////////
Anchor cables: ....

21

other floating storage, the single point mooring loading systems suffer the disadvantage of being unusable in severe weather, when dangers of spillage and problems of tanker handling increase.[36]

In the long term, pipelines must be used for production from large fields or groups of fields. Major problems associated with pipelines occur during the laying stages. Two basic offshore operations are involved. First, pipe sections are welded together on a lay barge from which the pipeline is lowered to the sea bed. Subsequently a trenching (bury) barge buries as great a length as possible. In laying operations especially, the difficulties are similar to those of exploration drilling. The barge must be kept in position by a suitable anchoring system, used also to pull it along as the pipeline is laid. Major problems are, however, presented by deep water. Until recently it has been virtually impossible to lay large diameter pipe (91 cm/36 in) in water deeper than about 180 m. Now a new generation of lay barges built to wide-beamed or semi-submersible designs is appearing. These can be used to lay pipelines in much deeper water, for example across the Norwegian Deep (Fig. 3.1)[37], which has hitherto been an insuperable barrier to linking the Norwegian fields to the coast of Norway. The job of burying pipelines is discussed below in connection with the characteristics of the sea bed itself.

A final set of problems posed by depth of water is that associated with underwater working. This is necessary at virtually all stages of offshore development, especially in test-drilling, pipeline laying and subsequent maintenance of pipelines, platforms, wellheads and other underwater equipment. At present by far the greater proportion of underwater work requiring direct working by men is carried out by divers, approximately 800 being currently employed in the North Sea operations. Mini-submarines (submersibles) of various designs are, however, being increasingly used, notably for inspection work, and would appear to have a promising future. By the late summer of 1974 out of 52 submersibles in operation around the world, 18 were employed by the oil industry, 8 of these in the North Sea.[38]

The chief environmental hazards in diving[39] are the high pressures and low temperatures encountered at depth, which pose serious difficulties even at present depths of operation down to 180 m. For prolonged working in depths greater than 60—90 m saturation diving techniques are necessary. These involve replacing air with a helium-oxygen mixture for breathing, as the nitrogen in air is poisonous when absorbed into the bloodstream under pressure at depth. The body of the diver has to absorb enough of the inert, non-poisonous helium to be in equilibrium with the pressure equivalent at the working depth. In order to avoid this absorbed gas being released into the bloodstream in bubble form when pressure is reduced as the diver ascends, thereby producing a serious disorder known as 'the bends' which can result in death, the diver requires slow decompression at a carefully regulated rate. At working depths in the order of 300 m, this rate is reckoned to be approximately one day for each 30 m of depth.[40] Techniques for controlling rates of compression and decompression are complex, and although the more immediate dangers are well-known, medical knowledge of their long-term effects on divers is still relatively limited. Additionally, in order to counteract low sea temperatures at depth, heat must be provided to maintain sufficiently high temperatures in the breathing supply and diving suit. This is generally done by means of a hot water supply.

Engineering hardware for diving is understandably complex in detail. Compression and decompression are generally carried out at the surface in a special decompression chamber on the deck of the rig, barge or other vessel. This chamber can be connected to a diving bell used to transport divers to and from the surface. Divers then work by leaving the bell for short periods at a time. Use of the bell means that compression can be maintained as long as necessary, relative to the work in hand. Already (December 1974) dives to 220 m (730 ft) have been carried out west of Shetland, a test dive to 260 m (850 ft) has been accomplished off the coast of West Africa,[41] and it is likely that the 300 m (1 000 ft) dive will soon be attained. However, the hazards are such that further development of subsea production systems and submersibles in which work can be done in a 'shirtsleeve environment' at normal atmospheric pressure will remain attractive.

B: The sea bed

Sea-bed characteristics, including soil materials, local submarine topography and bottom currents are also vital factors in offshore operations.[42] Apart from the obvious influence of water depth in the scaling-up of platform designs, the soils encountered are, next to wave conditions, the most important physical consideration in placing production platforms on the sea bed. The nature of the foundations influences the choice among steel, concrete or composite designs.[43] The materials composing the sea floor can be grouped into a variety of categories according to texture, mode of origin and other criteria (Fig. 3.10) and must be carefully studied at the design stage.[44] Major categories include sand and mud; clay; gravel, stones and other 'hard ground'; and solid rock. With the exception of solid rock, all are depositional—sand in particular occurs widely on top of both clay and rock. The broad pattern of distribution can be explained largely by reference to the mode of origin, particularly with respect to deposition during the several phases of the Pleistocene Ice Age (Fig. 3.4), followed by subsequent Holocene deposition and reworking of older deposits (Fig. 3.10).

During the Pleistocene large parts of the North Sea and other shallow seas around North West Europe were at various times above sea level, especially in the south, and often covered by ice originating in the adjacent high land masses of Scandinavia and the British Isles. Apart from a certain amount of topographic modification, the most notable example of which may be the Norwegian Deep,[45] the presence of ice has led to extensive deposition in the form of boulder clay and probably also moraines, as well as associated fluvio-glacial deposits. However, unlike glacial deposits on land, these have been modified considerably by marine action. Postulated moraines are commonly associated with deposits of gravel and stones,[46] notably coinciding with the centre of the East Shetland Basin, on the banks immediately west of the Norwegian Deep (Figs. 3.1 and 3.4). Boulder clay is probably the most widespread of all deposits, having been compacted into a 'basal conglomerate' in many places,[47] and often covered by sand and mud produced by reworking of older deposits (including fluvio-glacial material) and later Holocene deposition. Although it can withstand fairly substantial loading of the kind imposed by concrete bottom-supported platforms and is soft enough despite induration by ice to provide reasonable foundations for pile-supported structures, the problems associated with pipe-laying over boulder clay have not been satisfactorily solved to date.[48] It is a difficult material in which to bury pipelines.

Sand is particularly widespread in coastal areas, where fluvio-glacial streams have deposited, and present rivers continue to deposit, vast quantities. Some sand, especially in northern waters, is also

undoubtedly the result of reworking of glacial and fluvio-glacial deposits. Both sand and mud, also widespread in several areas, provide poor foundations for platforms. It seems fair to assume that these types of 'soft' bottom were among the factors causing delay in placing the first platform on the Frigg Field.[49] Also a special study has proved necessary to investigate the 'tricky' nature of the bottom in the area of the Thistle Field.[50] The principal hazard encountered so far with sand has been the difficulty in burying pipelines in a deposit which is most readily reworked by tidal and other bottom currents. These can lead to scour and exposure in an unpredictable manner of pipeline already buried. For example, no less than eight of the 43 miles of the line from the West Sole Field to Easington in Yorkshire were simultaneously exposed at various points along its length.[51] Linked to this type of scour is the formation of sand waves, which pose threats not only to platform foundations and pipelines, but also to navigation, by the alteration and infilling of shipping routes in shallow waters used by deep draught tankers.[52]

Near coasts on the northern part of the shelf, where post-glacial subsidence has led to the submergence of areas recently subjected to erosion, solid rock bottom is common, notably around the Northern Isles and Outer Hebrides. These relatively localised areas may prove to have a significance far beyond that suggested by their total extent, due to the use of the Northern Isles for pipeline landfalls (Figs. 6.1, 6.4). It is impossible to bury pipelines in solid rock without a great deal of blasting and infilling, and lines crossing this must consequently remain exposed to stresses imposed by tidal currents in relatively shallow water. Already, extensive operations have proved necessary a few miles off Firths Voe to level rock pinnacles in the way of the first Brent pipeline (Fig. 6.1).

3. The Coastline

In the context of the offshore oil industry, the coastline of Scotland[53] presents three distinct physical environments—high cliff, low rocky and low sandy or 'soft' coasts. The first type is almost invariably associated with a high degree of exposure to the open sea, notably along the east coast of the mainland, and western coasts of the Northern Isles, and is generally unsuited to development. The low rocky coasts, on the other hand, are common in areas in which glacial erosion has been pre-

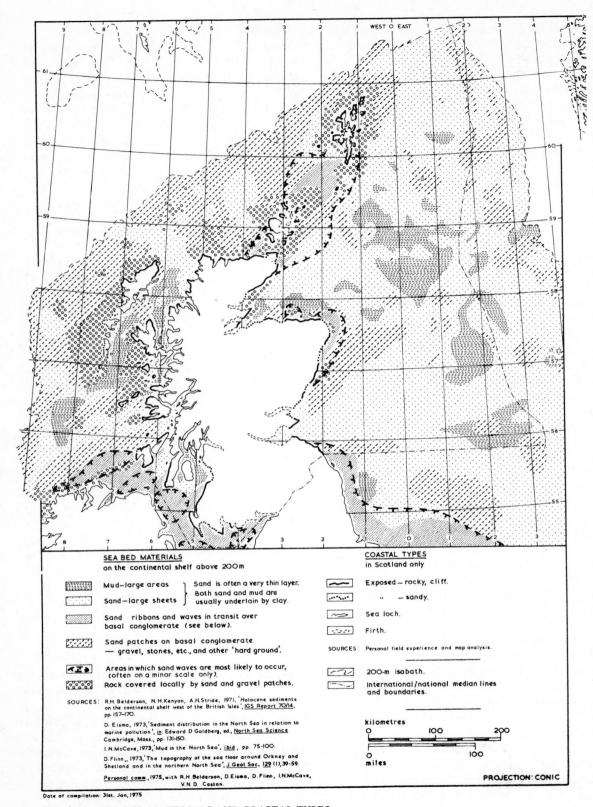

SEA BED MATERIALS
on the continental shelf above 200m

▦	Mud—large areas } Sand is often a very thin layer.
▫	Sand—large sheets } Both sand and mud are usually underlain by clay.
▨	Sand ribbons and waves in transit over basal conglomerate (see below).
▨	Sand patches on basal conglomerate — gravel, stones, etc., and other 'hard ground'.
◀▬▶	Areas in which sand waves are most likely to occur, (often on a minor scale only).
◉	Rock covered locally by sand and gravel patches.

SOURCES: R.H. Belderson, N.H.Kenyon, A.H.Stride, 1971, 'Holocene sediments on the continental shelf west of the British Isles', IGS Report 70/14, pp. 157-170.

D. Eisma, 1973, 'Sediment distribution in the North Sea in relation to marine pollution', in: Edward D Goldberg, ed., North Sea Science Cambridge, Mass., pp. 131-150.

I.N.McCave, 1973, 'Mud in the North Sea', ibid , pp. 75-100.

D.Flinn,, 1973, 'The topography of the sea floor around Orkney and Shetland and in the northern North Sea', J.Geol.Soc., 129 (1), 39-59.

Personal comm., 1975, with R.H. Belderson, D.Eisma, D.Flinn, I.N.McCave, V.N.D. Caston.

COASTAL TYPES
in Scotland only

⌇	Exposed — rocky, cliff.
⌇	" — sandy.
⌇	Sea loch.
⌇	Firth.

SOURCES: Personal field experience and map analysis.

⌇	200-m isobath.
⌇	International/national median lines and boundaries.

kilometres
0 ——— 100 ——— 200

miles
0 ——— 100

PROJECTION: CONIC

Date of compilation: 31st. Jan, 1975

FIG. 3.10 SEA-BED MATERIALS AND COASTAL TYPES

24

dominant, and which remain sheltered from the open sea. These coasts are often broken by relatively small sand and pebble beaches and estuaries, which sometimes provide good dry dock platform building sites when associated with easily excavable materials on beaches and adjacent areas. The leading example of this coastal type is the sea loch, so common along the west coast, Inner Hebrides and in Shetland. The key asset of the sea loch is the presence of sheltered deep water close inshore, necessary for building certain types of concrete platforms. Some of these must be floated in deep water during all construction phases. All must do so during the later stages of construction and have access to the open sea through a reasonably deep channel. Further, because of the deep water, the sea loch coast is also the first choice for tanker terminals, one of which already exists at Finnart on Loch Long for imported crude oil, while another is planned for Sullom Voe in Shetland. The suitability of sea loch coasts for the establishment of service bases, consequent upon their natural harbours, is being taken advantage of in Shetland (Fig. 6.1).

The low 'soft' coasts are of two kinds. One of these is the open sand or pebble beach characteristic of stretches of the east coast which, with isolated exceptions, would appear to be too exposed for development. Ground conditions are, however, often suitable for pipeline landfalls, as at Cruden Bay[54] and near St. Fergus (Fig. 6.6), and for excavation of platform dry docks where some shelter is provided by bars, for example Whiteness Head near Ardersier (Fig. 6.5). The second variety is the 'firth' type, developed in the Upper Palaeozoic rocks of the south-west, east and north, especially around mainland firths and to some extent in Orkney. These coasts are, with the exception of Orkney, characteristically estuarial, with combinations of sandy beaches, tidal mudflats, carselands (mainly raised estuarial shoreline deposits) and low rocky outcrops thickly mantled with glacial drift.[55] Large areas of flat land are available close to water which, if not naturally deep, can usually be dredged to provide suitable navigation channels. The thick boulder clay and estuarial sand and mud deposits are suitable for excavation of dry docks, for example at Nigg Bay (Fig. 6.5), and for port and terminal construction, for example on the Forth at Grangemouth and Hound Point. Associated flat land, very limited elsewhere in Scotland, provides good refinery locations, as at Grangemouth and potentially on the Cromarty Firth and Firth of Clyde (Figs. 6.5, 6.11, 6.16).

REFERENCES AND NOTES

1. STRIDE, A. H., CURRAY, H. J., MOORE, G. G., BELDERSON, R. H. (1969), 'Marine geology of the Atlantic continental margin of Europe.' *Phil. Trans. R. Soc. Ser. A.*, 264, pp. 31–75.
2. ANON (1974a), 'The amazing case of the absent-minded professors,' *The Oilman*, 14 December, pp. 8–9.
3. BOTT, M. H. P. (1974), 'Structure and evolution of the Shetland-Hebridean shelf, the Faroe block and the intervening region.' ROBERTS, D. G. (1974), 'Tectonic and stratigraphic evolution of the Rockall Plateau and Trough.' Both papers in *Petroleum and the Continental Shelf of North West Europe: the Geology and the Environment*, abstracts of contributed papers to a conference held under the auspices of the Institute of Petroleum, *et al.* Full proceedings to be published in 1975 by Applied Science Publishers, London.
4. *Ibid.*
5. BOTT, M. H. P., BROWITT, C. W. A., STACEY, A. P. (1971), 'The deep structure of the Iceland-Faroe Ridge,' *Marine Geophys. Res.*, 1(3), pp. 328–351.
6. ANON (1974a), *op. cit.*
7. ROBERTS, D. G. (1974), *op. cit.* EDEN, R. A. (1974), 'Explorers now looking closely at Atlantic margin,' *Petrol. Int.* 14(4), April, pp. 59–65.
8. See, for example, the abstracts of contributed papers cited in 3 above.
9. FOX, A. F. (1964), *The World of Oil*, Pergamon, Oxford, is a useful non-technical introduction to basic petroleum geology.
10. KENT, P. E. (1967), 'Outline geology of the southern North Sea Basin,' *Proc. Yorks. Geol. Soc.* 36(1), pp. 1–22.
11. DUNNING, F. W. (1970), *Geophysical Exploration*, HMSO (The Geological Museum) London, gives a detailed account of geophysical techniques.
12. BOTT, M. H. P., WATTS, A. B. (1970), 'Deep structure of the continental margin adjacent to the British Isles, *Inst. Geol. Sci. Report 70/14*, pp. 89–109.
13. HINDE, P. (1974), *The Exploration for Petroleum with particular reference to N.W. Europe*, Gas Council, London. Cazenove & Co. (1972), *The North Sea: the search for oil and gas and the implications for investment*. London, pp. 69–72. EDEN, R. A. (1974), *op. cit.*
14. Abstracts of contributed papers, cited in 3 above (papers 12–19, 23).
15. *Ibid.*, papers 20, 21, 28, 33, 34, 37, 38.

16. *Ibid.,* paper 27.

    BUCKMAN, D., CRANFIELD, J. (1974), 'North Sea-score card is filling up fast,' *Petrol. Int.,* 14(11), November, p. 20: Generalised North Sea stratigraphy.

17. WHITBREAD, D. R. (1974), 'Geology and petroleum possibilities west of UK' from the abstracts of contributed papers, cited in 3 above.

    BINNS, P. E., McQUILLIN, R., FANNIN, N. G. T., KENOLTY, N. ARDUS, D. A. (1974), 'Structure and stratigraphy of sedimentary basins in the Sea of the Hebrides and Minches,' *ibid.*

    BOTT, M. H. P. (1974) op. cit.

18. HAMILTON, A. (1974) 'North Sea Oil Review' the missing ingredient in the West Shetlands,' *The Financial Times,* 6 December.

19. STRIDE, A. H. *et. al.,* (1969) *op. cit:* EDEN, R. A. (1974), *op. cit.*

20. ALLCOCK, L. C. (1974), 'Offshore oil and gas: the future engineering contribution,' *Inst. Mech. Engrs. Proc.,* 188(2). An edited version appears in *The Chartered Mechanical Engineer,* 21(1), pp. 47–51. This is a general review of the scope of offshore engineering.

21. HOHN, R. (1973), 'On the climatology of the North Sea,' in E. D. GOLDBERG (ed.), *North Sea Science,* M.I.T. Press, Cambridge, Mass., pp. 183–236.

22. CAZENOVE & Co., (1972), *op. cit.,* pp. 37–40.

23. OGDEN, R. G. (1970), 'Weather forecasting for the North Sea,' in K. D. TROUP (ed.), *Norspec 70,* Thomas Reed, London, pp. 77–87.

24. DRAPER, L. (1970), 'North Sea wave data,' *ibid,* pp. 243–8

25. CHILDERS, M. A. (1974), 'Deep water mooring Part I: Environmental factors control station-keeping methods,' *Petrol. Engr.,* September, pp. 36–58.

26. COTTRILL, A. (1974), 'Gravity platforms—who is proposing what,' *New Civil Engineering Special Review: North Sea Oil, May,* pp. 39–44.

    McPHEE, W. S., REEVES, S. J. (1974), 'Towards ever-deeper platforms,' *New Civil Engineer,* Special Supplement: *Offshore Structures,* September, pp. 91–8.

27. BREWER, J. H., SHRUM, S. J. (1973), 'Tension-leg platform will get at-sea test next year,' *Oil Gas J.,* October 8, pp. 88–91.

    McDONALD, R. D. (1974), 'The design and field testing of the "Triton" tension-leg fixed platform and its future application for petroleum production and processing in deep water,' Paper No. OTC. 2104, *Offshore Technology Conference,* Dallas.

28. WALKER, R. C. (1973) 'Problems in getting the oil ashore,' in *North Sea oil—the challenge and the implications,* Heriot-Watt University Lectures, 1973, Edinburgh, pp. 15–25

29. Department of Energy (1974), *Guidance on the Design and Construction of Offshore Installations.* HMSO, London, Section 2: 'Environmental considerations,' pp. 3–22.

30. WALKER, R. C. (1973), *op. cit.*

31. TUBB, M. (1974) 'A new platform design for deep-water drilling,' *Ocean Industry,* May 1974.

32. ANON (1974b), 'Subsea systems are in worldwide use,' *Offshore,* 34(9) August pp. 71–3.

    ANON (1974c), 'North Sea SPM loading feasible but not acceptable for major field,' *Petrol. Int.,* 14(11), December, pp. 46–8, contains details of operation of Ekofisk.

33. ANON (1974d), 'Production activity builds up fast on the North Sea,' *Petrol. Int.,* 14(9), September, 28–32.

34. ALLCOCK, L. C. (1974), *op. cit.*

35. BAX, J. D. (1974), 'First SPAR storage-loading buoy project posed complex design, construction problem, *Oil Gas J.,* 10 June pp. 53–7.

36. ANON (1974c), article on details of operation of Ekofisk, cited in 32 above.

37. O'DONNELL, J. P. (1974), 'Pipeline construction in North Sea waters reflects area's boom,' *Offshore,* 34(8), pp. 57–67.

38. ANON (1974b), *op. cit.* p. 82.

39. HUGHES, D. M. (1974), 'What is it like to work at 1 000 ft?' *ibid,* pp. 67–70.

    FLEMING, N. C. (1968), 'Diving technology,' in *Offshore Europe,* first edition, Scientific Surveys (Offshore) Ltd., London, pp. 91–101.

40. HUGHES, D. M. (1974), *op. cit.*

41. ANON (1974e), *The Oilman,* 14 December 1974.

42. CASTON, V. N. D. (1974), 'Bathymetry of the Northern North Sea: knowledge is vital for offshore oil,' *Offshore,* 34(2), pp. 76–84.

43. McFEETERS, J. N. (1973), 'Fundamental considerations regarding bottom supported and pile supported drilling and production platforms in the North Sea,' *Offshore Scotland Conference,* Day 2, p. R1. Offshore Services, London.

44. STUBBS, S. B. (1974), 'Seabed considerations—putting gravity on a firm footing,' *New Civil Engineer Special Supplement: Offshore Structures,* September, pp. 40–5.

45. HOLTEDAHL, O. (1970), 'On the morphology of the West Greenland shelf with general remarks on the "marginal channel" problem,' *Marine Geol.,* 8, pp. 155–72.

46. EISMA, D. (1973), 'Sediment distribution in the North Sea in relation to marine pollution,' in *North Sea Science* cited in 21 above, pp. 131–50.

47. BELDERSON, R. H., KENYON, N. H., STRIDE, A. H. (1971), 'Holocene sediments on the continental shelf west of the British

Isles,' *Inst. Geol. Sci.*, Report 70/14, pp. 157–70.

48. WALKER, R. C. (1973), *op. cit.*

49. ANON (1974f), *Petroleum Times,* 78 (1988), p. 4.

50. BUCKMAN, D. (1974), 'Foul winter weather fails to slow North Sea's run of success,' *Petrol. Int.,* 14 (4), April, pp. 28–35.

51. ROSS, E. A. (1970), 'An introduction to North Sea offshore exploration,' in *Norspec 70,* cited in 23 above, pp. 32–4.

52. CLOET, R. L. (1970), 'How deep is the sea?' *ibid*, pp. 109–14.

53. STEERS, J. A. (1973), *The Coastline of Scotland,* Cambridge University Press, is a detailed descriptive account.

54. RITCHIE, W. (1973), 'Piping in the new wealth,' *The Geogr. Mag.,* 45(11), pp. 777–9.

55. SISSONS, J. B. (1967), *The Evolution of Scotland's Scenery*, Oliver & Boyd, Edinburgh, pp. 170–89 provides a detailed introduction to the Forth and Tay examples.

4 Offshore—Licensing and Discoveries

Discoveries in recent years and their subsequent evaluation have firmly established the North Sea as a major oil and gas province. Although they make relatively little difference to world reserves overall (p. 3), their location in Western Europe has greatly enhanced their strategic significance, especially in view of recent changes in the world price and availability of oil.

1. The Legal Framework

Exploration for oil and gas first necessitated creating a legal framework to cover the sea areas adjacent to the countries bordering the North Sea. The question of territorial control was determined by the Geneva Convention of 1958 on the Law of the Sea,[1] which gave bordering states sovereign rights over the natural resources below the sea bed of the continental shelf to a depth of 200 m. Subsequently, the division of the North Sea was agreed between governments, mainly on the basis of median lines, that is, the principle of equal distance from their respective coasts. Still unresolved, however, are the boundaries between England, France and Belgium, and a short section of the median line between Scotland and Norway (see below).

The Geneva Convention was ratified by the UK Government in May 1974, after the enactment of the Continental Shelf Act in April 1964. The latter empowers the Secretary of State for Energy on behalf of the Crown to grant licences to search for and obtain oil and gas in designated areas of the Continental Shelf; these areas being defined by the Continental Shelf (Designation of Areas) Order 1964, and subsequent Continental Shelf (Designation of Additional Areas) Orders in 1965, 1968 and 1971.

Under the Continental Shelf (Jurisdiction) Orders 1964 and 1965, both subsequently superseded by the Continental Shelf (Jurisdiction) Order 1968,[2] the UK sector of the North Sea was subdivided into Scottish and English areas along latitude 55°50'N (the latitude of the border at Berwick), thereby recognising the extension of Scottish and English law over the sea areas of the respective countries east to the median line. The Continental Shelf (Jurisdiction) Order 1968 also extends the jurisdiction of English, Northern Irish and Scottish law over the Irish and adjacent sea areas. The 1968 Order is an extension of the maritime jurisdiction earlier defined in the Wireless Telegraphy Act 1949, the Radioactive Substances Act 1960, and subsequently by the Mineral Workings (Offshore Installations) Act 1971.

The limits of the Scottish area are indicated in

TABLE 4.1   NORTH SEA–APPROXIMATE SIZE OF AREAS UNDER THE JURISDICTION OF BORDERING COUNTRIES

		Approximate Area		Percentage of Total Area
		km²	ml²	
United Kingdom	Scotland	160 000	62 500	30
	England	84 000	32 800	16
Norway		131 000	51 200	25
Netherlands		55 800	21 800	$10\frac{1}{2}$
Denmark		48 000	18 750	$9\frac{1}{2}$
West Germany		35 600	13 900	7
Belgium		4 000	1 600	1
France		4 000	1 600	1
North Sea		525 000	204 150	100

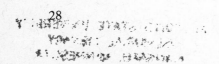

Fig. 4.5. Any extension beyond the present northern limit (62°N) will be subject to further negotiations with the Norwegian and Faeroese governments. The median line in the east is agreed except for a small section north of 61°44'12". No limit has been agreed so far to the west of the country further north than 55°20'N, 6°00'W (off the Mull of Kintyre). In 1974, however, the UK Government laid claim to 137 000 km² (52 000 ml²) of the sea bed around Rockall. Since Rockall was incorporated into Scotland several years ago the claim, if successful, will bring this area under Scottish jurisdiction. Although neither Eire nor Faroe have yet recognised this claim, its strength is worth considering in the light of the bathymetry of the sea bed west of the British Isles, and the respective distances between Rockall and the Irish, Faroese and Scottish (Western Isles including St. Kilda) coasts (Fig. 3.1).

Article 6 of the Geneva Convention defined the areal extent of the various North Sea sectors. This was subsequently modified by the extension of the West German sector west to the median line in 1970 following agreements with the Netherlands and Denmark. The unevenness of the line dividing the Danish and West German sectors is explained by the fact that the Danes retained the area in which they had already undertaken exploration (Fig. 4.4).

2. Offshore Concession Areas and Licensing

For licensing purposes each state, except Denmark, has superimposed a geographical grid on its sector (Fig. 4.1). All use a graticule drawn at one degree intervals of latitude and longitude as their primary system of division, but they have different systems of sub-division. The size of individual blocks decreases from south to north in each system because lines of longitude converge polewards, thus UK blocks vary in size from 210–240 km², Dutch blocks from 390–420 km², and Norwegian ones from 500–570 km². Part blocks adjacent to median lines or coasts may be either larger or smaller than these figures and licences may comprise one or more blocks or parts thereof. It will be appreciated that for ease of cartographical representation maps of the North Sea most commonly project blocks as rectangles within a particular sector.

UK offshore licences are of two types. Exploration licences (application fee £20) enable the holder to undertake exploratory work, including seismic survey and exploratory drilling, in sea areas not covered by production licences. Duration of the licence is three years at an annual rental of £1 000. Production licences (application fee £200) are granted for specific blocks and give the holder exclusive rights to search for and 'get' oil and gas. An additional fee of £5 per block is payable by holders of more than ten blocks. Since the North Sea was still largely an unproven area, licensing in the first and succeeding rounds between 1964 and 1972 was designed to encourage the greatest number of applications, and so make for as rapid and thorough exploration of resources as possible. A break with this procedure, however, occurred in 1971 when fifteen blocks (thirteen in the Scottish area) were auctioned—the final application date for sealed cash tenders was 20th August 1971. The highest bid tendered for a single block (Block 211/21) was £21.05 million by Shell Expro.[3]

Production licences are initially granted for six years. During each of the first two rounds (1964 and 1965), licensees were required to pay a rental of £25 per square kilometre for the first and each succeeding year, in addition to the application fee mentioned above. In the third and fourth rounds (1969–70 and 1971–72), the rental was increased to £30 and £45 respectively. Subject to satisfactory compliance with the terms and conditions under which the licence was granted, including completion of an agreed programme of work, the licence may be continued for a further forty years subject to surrender of one-half of the area held and an increase in annual rental. For first and second round licensees, this rental commences at £40 in the seventh year and increases by £25 per annum to a maximum of £290 in the seventeenth and subsequent years. Rentals for third and fourth round licensees begin at £50 in the seventh year and increase by £30 per annum to a maximum of £350 in the seventeenth and subsequent years. Rental payments are deductable from royalty (see (ii) p. 30 below).

Onshore production licences differ from offshore production licences in terms of annual rental and the maximum size (500 km²) of licence area allowed. Also they may be applied for at any time, unlike the applications invited by the Government at specific times for offshore licences. Onshore production licences are required to search for and 'get' oil and gas on land and in coastal sea areas classified as 'landward' (Fig. 4.5).

The offshore licence surrender system was designed to encourage exploration and to avoid territory being held without reasonable activity. Surrendered territory may be re-offered for licensing. For example, 77% of territory licensed in 1964 was surrendered in 1970. 44% of this surrendered territory was re-offered under the Fourth Round allocation and 16% was re-allocated.

Offshore production licensees are also required to:

(i) Inform the Minister (Secretary of State for Energy) of any change in rig position, change in the consortium or other material facts.
(ii) Pay a royalty of $12\frac{1}{2}\%$ of the well-head value of any petroleum produced. (The well-head value is derived by deducting from the landed value reasonable costs involved in delivering the oil and gas ashore.)

(iii) Ensure that any petroleum discovered shall normally be brought to the United Kingdom.
(iv) Ensure that any natural gas discovered which is to be sold as fuel is offered in the first instance to the British Gas Corporation.
(v) Fulfil certain minimum work obligations, the terms of which are negotiated for each licence.

At present exploration and production operations are governed by the statutory regulations made

TABLE 4.2   CONTINENTAL SHELF – SUMMARY OF LICENSING ACTIVITIES, 1963–74

1. To 31 December 1964

1963	Danish sector allocated in October, to A. P. Møller Group of companies. Dansk Undergrunds Consortium (DUC) formed.
1964	First Round of licences allocated in September in UK continental shelf (North Sea): English area – 262 blocks; Scottish area – 86 blocks.
1964	West German sector allocated to the West German North Sea Consortium.

2. 1 January – 31 December 1965

1965	First Round of licences allocated in August in Norwegian sector – 78 blocks.
1965	West German North Sea Consortium relinquish part of holdings in West German sector.
1965	Second Round of licences allocated in November in UK continental shelf (North Sea and Irish Sea): English area – 45 blocks; Scottish area – 82 blocks.

3. 1 January 1966 – 30 June 1970

1967	West German North Sea Consortium relinquish another part of holdings in West German sector.
1968	First Round of licences allocated in March in Dutch sector – 102 blocks.
1969	Second Round of licences allocated in October in Norwegian sector – 13 blocks.
1969–70	Third Round of licences allocated in September in UK continental shelf (North Sea, west of Orkney and Shetland, Irish and Celtic Seas): English area – 35 blocks; Scottish area – 71 blocks. Statutory relinquishment of 50% of First Round licence areas required in June 1970. Actually 75% surrendered, mostly in northern English waters outside the productive gas belt in the southern North Sea.

4. 1 July 1970 – 31 August 1972

1970	Second Round of licences allocated in September in Dutch sector – 34 blocks.
1970	Agreements reached in October between West German, Danish and Dutch governments allowing extension of West German sector to mid-North Sea median line. West Germany recognised concession rights previously issued by Denmark and the Netherlands, but the West German North Sea Consortium were granted exploration rights in areas previously unexplored in the newly acquired area.
1971	34% of First Round licences surrendered in Norwegian sector.
1971–72	Allocation of Fourth Round licences begun in August 1971 in UK continental shelf (North Sea, west of Orkney and Shetland, Irish and Celtic Seas) completed in March 1972: English area – 66 blocks (North Sea – 23, including 2 by tender, Irish and Celtic Seas – 43); Scottish area – 216 blocks (North Sea – 158, including 13 by tender, west of Orkney and Shetland – 58). Statutory relinquishment of 50% of Second Round licences in November 1971.
1972	In February, A. P. Møller Group obtained two year extension to initial exploration period of their concession in Danish sector.
1972	In June, Norwegian Government announced its intention to invite applications for a new issue of blocks in the autumn. In fact, no invitation was issued that year.

5. 1 September 1972 – 31 December 1974

1973–74	Formal invitation issued in July 1973 by Norwegian Government for Third Round of licences for 32 blocks (11 440 km² 4 450 ml²). 47 applications for 26 blocks received by September from companies and groups totalling 175 companies. Third Round of licences allocated in November 1974 in Norwegian sector – 8 blocks or part blocks (2 328 km² 909 ml²) to five company groups, in each of which the Norwegian Statoil company has a minimum 50% participation. 6 blocks (29/9, 30/7, 24/9, 15/11, 15/12, 6/3) close to median line, suggesting prospects of extension into the Norwegian sector of oil and gas reserves already discovered in the Scottish fields of Alwyn, Total 3/15, Total 3/19, Beryl and Maureen respectively.
1973–74	Fifth Round of licensing in UK continental shelf withheld until government review of licensing terms completed.
1974	Dutch Government grant several production licences to speed up development of commercially proven offshore gas finds, in response to realisation that onshore gas fields no longer sufficient to meet consumption forecasts.

under the Petroleum (Production) Act 1934 and the Continental Shelf Act 1964. In July 1974 the UK Government announced that it intended to take action under five heads to ensure a greater government share of the profits from oil exploitation and greater control over developments:

(i) An additional tax will be imposed on companies' profits from the continental shelf and various loop-holes in the rules on existing taxation of their profits will be closed.
(ii) It will be made a condition of future licences that the licensees shall, if the government so requires, grant majority participation to the state in all fields discovered under these licences.
(iii) Companies will be invited to enter into discussions with the government about majority state partici-pation in existing licences for commercial fields.
(iv) A British National Oil Corporation will be set up through which the government will exercise its participation rights. The headquarters of the BNOC will be in Scotland.
(v) The government will extend its powers of physical control over offshore operations, including pro-duction, and over pipeline developments.

A comprehensive review of the terms under which licences are to be issued in the future is being made by the Department of Energy. It is unlikely that any future licensing will take place until this review has been completed.

Under the terms of the licences awarded in the Fourth Round allocation of the UK continental shelf (282 blocks, 118 licences, 213 companies), the agreed minimum work programme involves seismic investigation on an extensive scale and requires at least 224 explora-tory wells to be drilled within the six years' initial period of the licences. This compares with an agreed minimum of 189 wells for the first three rounds of licensing. By the autumn of 1974 a substantial part of the drilling commitments—nearly 190 wells—remained to be ful-filled, and most of those were entered into during the Fourth Round.

The Scottish area of the North Sea has enjoyed a comparatively high ratio of success in drilling. This is particularly so if the study is limited to what are reckoned by the industry to be commercially viable finds under present conditions, that is, fields capable of sustained growth in excess of 30 000 barrels of oil per day, or say about 3 000 000 m³ of gas per day. Exclud-ing the Scottish part of the Frigg gas field, the overall success ratio is 1:8.[4]

3. Time Scale of Licensing Activity

A summary of continental shelf offshore licensing is given in Fig. 4.2. Figs. 4.1—4.4 illustrate how licensing proceeded, together with discoveries made during the period 1963—72. Apart from the statutory relinquish-ment of certain blocks or part blocks, and the alloca-tion of a further eight blocks in the Norwegian sector in 1974 (see below), concession areas at the end of 1974 differ little from the position portrayed in Fig. 4.4. Additional detail for the period 1972—74 has, however, been included for Scottish waters (Figs. 4.5, 4.6). A complete list of North Sea discoveries and their distribution on the basis of jurisdiction areas is given in Tables 4.5 and 4.6.

It is worth noting, however, that because Figs. 4.1—4.4 are drawn primarily to show the allocation of blocks in successive licensing rounds, the maps cover variable periods of time. Care should there-fore be taken when studying the rate at which discover-ies have been made.

The total number of blocks licensed in Scottish waters during the first four rounds was 455 (cf. English waters 408). Scottish blocks were distributed as follows: North Sea — 360; west of Orkney and Shetland — 94; Irish Sea — 1. The total figure is considerably less than the number of blocks offered, since in each allocation certain blocks remained unallocated. On completion of the Fourth Round, 347 companies were operating in the Scottish offshore area.[5]

Total receipts from offshore licences allocated during the first four rounds between 1st April 1964 and 30th September 1972 within the UK continental shelf were:

Exploration Licences — £188 600; Production Licences — £47 389 606 (includes £37 213 654 for the 15 blocks offered for tender in August 1971; Royalties — £18 382 618; Total — £65 960 824.[6]

4. Distribution of Oil and Gas Fields

Interest in the North Sea as a potentially rich source of hydro-carbons gained momentum following the discovery of the onshore Slochteren gas field in 1959

(Fig. 4.1). UK Government encouragement of exploration in English waters was rewarded by considerable finds of gas from the mid-sixties onwards. The largest field, Leman, occurs in deposits similar to those which form the reservoir for the onshore Groningen fields. However, it was the Dansk Undergrunds Consortium's first oil strike in 1967 — a discovery which later proved to be only marginally economic — and the subsequent commercial finds in December 1969 in Scottish waters (Montrose — Block 22/18), and Norwegian waters (Ekofisk — Block 2/4), that firmly focused world attention on the North Sea. Further encouraging discoveries in Norwegian and Scottish waters in 1970 and 1971 (Fig. 4.4) were reflected in the large sums which the oil companies were prepared to pay for licences for the blocks offered for tender during the Fourth Round by the UK Government in August 1971 (p. 29).

The increase in drilling activity in Scottish waters from September 1972 is illustrated by Table 4.3. The remarkable rate of success in operations in the East Shetland Basin maintained interest there at a high level over the two years ending December 1974, although an increasing number of rigs were attracted to the Central North Sea Basin following some encouraging finds made there during the spring and summer of 1974. On average in 1974 between 60% and 70% of all rigs operating in the North Sea were in Scottish waters. This is borne out by reference to Table 4.4 which shows the distribution of rigs in November 1974. Lately, most rigs in Scottish waters have been drilling on or near existing fields to evaluate reserves and establish extensions.

Initially, drilling in the northern North Sea was handicapped by having to depend on rigs which were not specifically designed to cope with the rigorous conditions experienced, particularly in winter. However, today a new generation of larger semi-submersibles has been designed and is being brought into service (p. 18). The result is that there is no longer a marked seasonal reduction in winter in the number of rigs in operation (Table 4.3).

Table 4.5 lists North Sea oil and gas discoveries to the end of 1974. Although comparison with Fig. 4.4 shows that there have been a number of notable discoveries outside Scottish waters in the period 1972–74, especially in the Norwegian sector, activity elsewhere has generally been much more restricted, and is one of the reasons why rigs have been concentrated in Scottish waters, as indicated in Table 4.4. Government policies, notably in Norway, have favoured a more gradual rate of development. In the Norwegian case the small size of the domestic market and the desire to minimise the inflationary effects of too rapid a rate of development

TABLE 4.3   LOCATION AND TYPES OF RIGS OPERATING IN SCOTTISH WATERS, SEPTEMBER 1972 – DECEMBER 1974

Location	Sept. '72	Dec. '72	Feb. '73	Apr. '73	Jul. '73	Oct. '73	Dec. '73	Mar. '74	Jul. '74	Oct. '74	Nov. '74	Dec. '74
East Shetland Basin	2	2	2	5	10	10	8	10	14	12	12	21
West of Shetland and Orkney	–	–	–	–	–	–	–	1[a]	2[b]	2[c]	3[d]	–
Central North Sea	4	5	5	5	3	4	6	9	7	9	10	6
Total	6	7	7	10	13	14	14	20	23	23	25	27
Type												
Semi-submersible	2	7	7	8	11	14	14	20	21	23	23	26
Drillship	4	–	–	2	1	–	–	–	2	–	1	–
Jack-up	–	–	–	–	1	–	–	–	–	–	1	1
Total	6	7	7	10	13	14	14	20	23	23	25	27

a. West of Shetland
b. West of Shetland
c. One to west of Shetland, one to west of Orkney
d. Two to west of Shetland, one to west of Orkney

Sources: *Standing Conference on North Sea Oil*
Petroleum Times
The Oilman

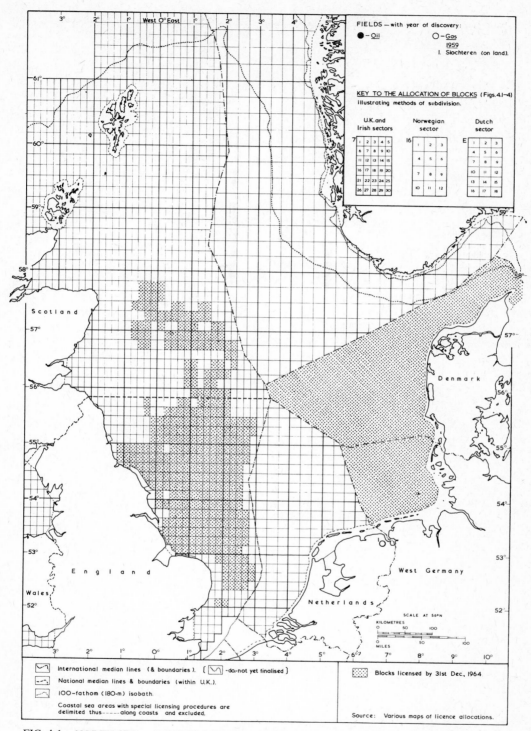

FIELDS — with year of discovery:
● – Oil   ○ – Gas
                      1959
              I. Slochteren (on land).

KEY TO THE ALLOCATION OF BLOCKS (Figs. 4.1–4)
illustrating methods of subdivision.

U.K. and Irish sectors	Norwegian sector	Dutch sector

Scotland

England

Wales

Netherlands

Denmark

West Germany

SCALE AT 56°N
KILOMETRES
0   50   100
MILES
0   50   100

International median lines (& boundaries). [-do- not yet finalised]

National median lines & boundaries (within U.K.).

100-fathom (180-m) isobath.

Coastal sea areas with special licensing procedures are
delimited thus ----- along coasts and excluded.

Blocks licensed by 31st Dec, 1964

Source: Various maps of licence allocations.

FIG. 4.1   NORTH SEA – ALLOCATION OF BLOCKS TO 31ST DECEMBER 1964

33

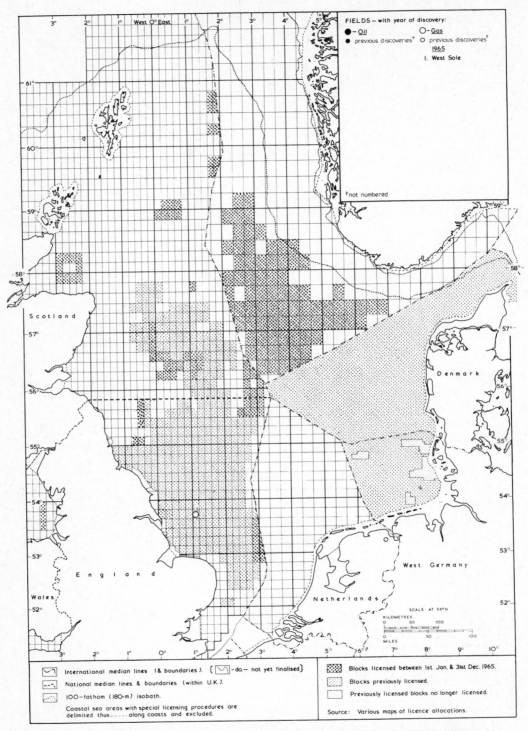

FIG. 4.2   NORTH SEA – ALLOCATION OF BLOCKS, 1ST JANUARY – 31ST DECEMBER 1965

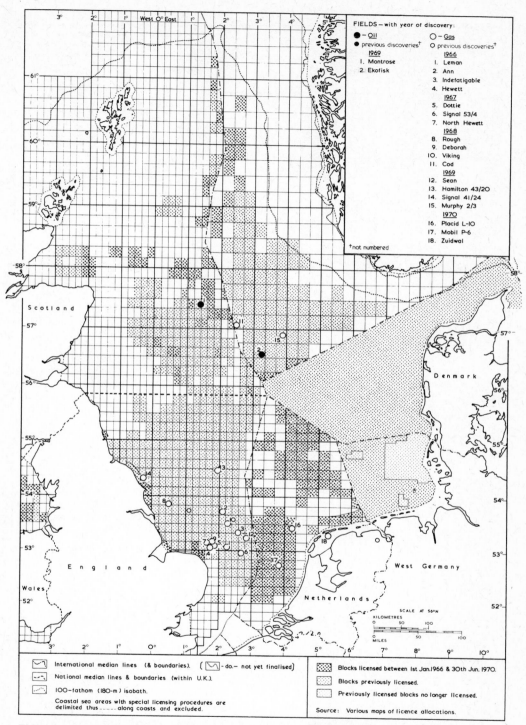

FIELDS—with year of discovery:
● —Oil ○ —Gas
● previous discoveries† ○ previous discoveries†

1969
1. Montrose
2. Ekofisk

1966
1. Leman
2. Ann
3. Indefatigable
4. Hewett
1967
5. Dottie
6. Signal 53/4
7. North Hewett
1968
8. Rough
9. Deborah
10. Viking
11. Cod
1969
12. Sean
13. Hamilton 43/20
14. Signal 41/24
15. Murphy 2/3
1970
16. Placid L-10
17. Mobil P-6
18. Zuidwal

†not numbered

SCALE AT 56°N
KILOMETRES   50   100
MILES   50   100

International median lines (& boundaries). (⌐⌐ - do.- not yet finalised)
National median lines & boundaries (within U.K.).
100-fathom (180-m) isobath.
Coastal sea areas with special licensing procedures are delimited thus ----- along coasts and excluded.

Blocks licensed between 1st Jan.1966 & 30th Jun. 1970.
Blocks previously licensed.
Previously licensed blocks no longer licensed.

Source: Various maps of licence allocations.

FIG. 4.3   NORTH SEA – ALLOCATION OF BLOCKS, 1ST JANUARY 1966 – 30TH JUNE 1970

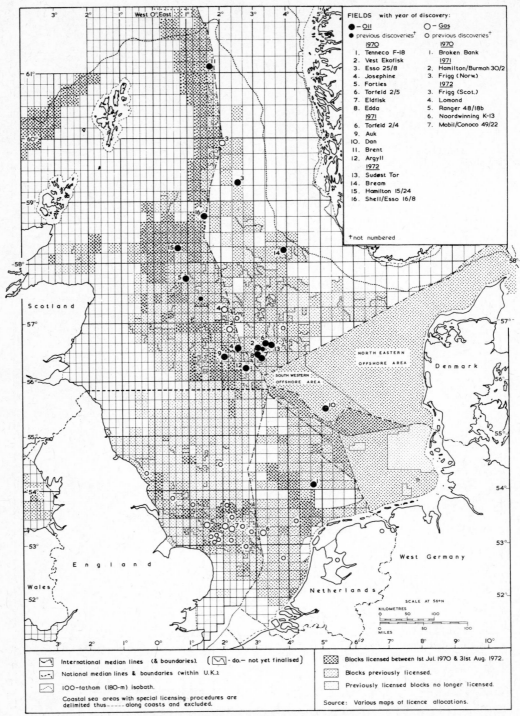

FIELDS with year of discovery:

● – Oil   ○ – Gas
◓ previous discoveries†   ◑ previous discoveries†

1970
1. Tenneco F-18
2. Vest Ekofisk
3. Esso 25/8
4. Josephine
5. Forties
6. Torfeld 2/5
7. Eldfisk
8. Edda
1971
6. Torfeld 2/4
9. Auk
10. Dan
11. Brent
12. Argyll
1972
13. Sudest Tor
14. Bream
15. Hamilton 15/24
16. Shell/Esso 16/8

1970
1. Broken Bank
1971
2. Hamilton/Burmah 30/2
3. Frigg (Norw.)
1972
3. Frigg (Scot.)
4. Lomond
5. Ranger 48/18b
6. Noordwinning K-13
7. Mobil/Conoco 49/22

†not numbered

International median lines (& boundaries). (⌐⌐ - do.– not yet finalised)

National median lines & boundaries (within U.K.)

100-fathom (180-m) isobath.

Coastal sea areas with special licensing procedures are delimited thus------along coasts and excluded.

▨ Blocks licensed between 1st Jul. 1970 & 31st Aug. 1972.

░ Blocks previously licensed.

☐ Previously licensed blocks no longer licensed.

Source: Various maps of licence allocations.

FIG. 4.4   NORTH SEA – ALLOCATION OF BLOCKS, 1ST JULY 1970 – 31ST AUGUST 1972

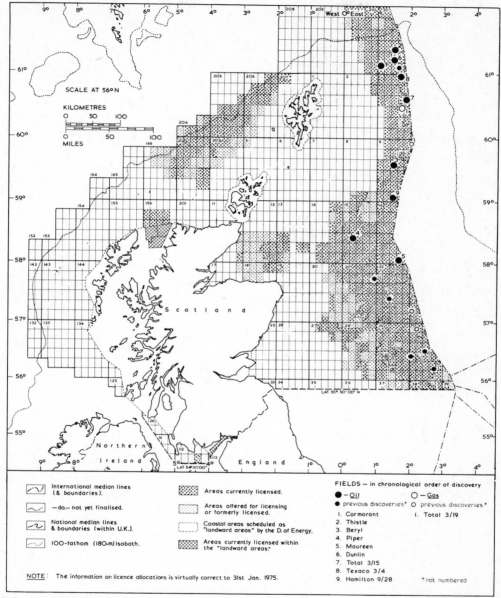

NOTE: The information on licence allocations is virtually correct to 31st Jan. 1975.

FIG. 4.5   SCOTTISH WATERS – CONCESSION AREAS AND DISCOVERIES, 1ST SEPTEMBER 1972 –
31ST AUGUST 1973

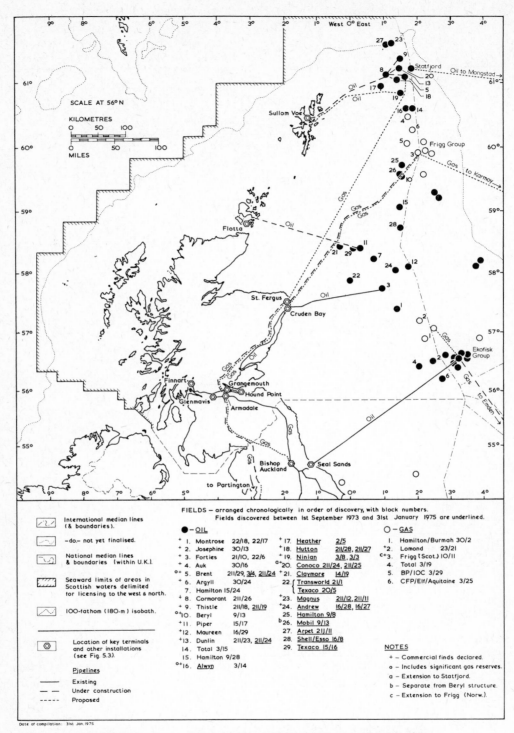

FIELDS – arranged chronologically in order of discovery, with block numbers.
Fields discovered between 1st September 1973 and 31st January 1975 are underlined.

International median lines (& boundaries).

–do.– not yet finalised.

National median lines & boundaries (within U.K.).

Seaward limits of areas in Scottish waters delimited for licensing to the west & north.

100-fathom (180-m) isobath.

Location of key terminals and other installations (see Fig. 5.3).

Pipelines

Existing

Under construction

Proposed

● – OIL

+	1.	Montrose	22/18, 22/17
+	2.	Josephine	30/13
+	3.	Forties	21/10, 22/6
+	4.	Auk	30/16
°+	5.	Brent	211/29, 3/4, 211/24
+	6.	Argyll	30/24
	7.	Hamilton	15/24
+	8.	Cormorant	211/26
+	9.	Thistle	211/18, 211/19
°+	10.	Beryl	9/13
+	11.	Piper	15/17
+	12.	Maureen	16/29
+	13.	Dunlin	211/23, 211/24
	14.	Total	3/15
	15.	Hamilton	9/28
°+	16.	Alwyn	3/14

+	17.	Heather	2/5
+	18.	Hutton	211/28, 211/27
+	19.	Ninian	3/8, 3/3
a+	20.	Conoco	211/24, 211/25
+	21.	Claymore	14/19
	22.	Transworld	21/1
		Texaco	20/5
+	23.	Magnus	211/12, 211/11
+	24.	Andrew	16/28, 16/27
	25.	Hamilton	9/8
b	26.	Mobil	9/13
	27.	Arpet	211/1
	28.	Shell/Esso	16/8
	29.	Texaco	15/16

○ – GAS

1. Hamilton/Burmah 30/2
+2. Lomond 23/21
c+3. Frigg (Scot.) 10/11
4. Total 3/19
5. BP/IOC 3/29
6. CFP/Elf/Aquitaine 3/25

NOTES

+ – Commercial finds declared.
o – Includes significant gas reserves.
a – Extension to Statfjord.
b – Separate from Beryl structure.
c – Extension to Frigg (Norw.).

Date of compilation: 31st Jan. 1975

FIG. 4.6   SCOTTISH WATERS – FIELDS DISCOVERED TO 31ST JANUARY 1975

on the national economy have had a strong influence in determining government policy towards the petroleum industry. In the Netherlands gas supplies from onshore fields have been sufficient to meet domestic needs. However, since 1973, there has been a growing realisation that onshore gas supplies will soon require to be supplemented by offshore supplies. This is the reason why several production licences were granted in 1974 to speed up the development of commercial finds offshore (Table 4.2). There are plans to bring

TABLE 4.4　CONTINENTAL SHELF – LOCATION OF RIGS IN TERMS OF JURISDICTION AREAS, NOVEMBER 1974

Northern North Sea	Scottish Waters	22	West of Shetland and Orkney	Scottish Waters	3
	Norwegian Waters	5			
	Sub-total	27			
			Irish Sea	English Waters	1
Southern North Sea	English Waters	1			
	Danish Waters	5	Celtic Sea	Irish Waters	2
	Dutch Waters	1			
	German Waters	2	Baltic Sea	Swedish Waters	1
	Sub-total	9			
North Sea		36	Other		7

Continental Shelf TOTAL 43

Source: *Petroleum Times*

TABLE 4.5　NORTH SEA – OIL AND GAS DISCOVERIES, WITH BLOCK NUMBERS, TO 31 JANUARY 1975
Discoveries are listed from north to south. Commercial finds proven to date are shown in italics

OIL

Scottish Waters

1. Arpet 211/11
2. *Magnus* 211/12, 211/17
3. *Thistle* 211/18, 211/19
4. *Dunlin* 211/23, 211/24
5. Conoco 211/24, 211/25
6. *Cormorant* 211/26
7. *Hutton* 211/28, 211/27
○8. *Brent* 211/29, 3/4, 211/24
9. *Heather* 2/5
10. *Ninian* 3/8, 3/3
○11. *Alwyn* 3/14
12. Total 3/15
○13. Hamilton 9/8
○14. *Beryl* 9/13
○15. Mobil 9/13
16. Hamilton 9/28
17. Shell 16/8

Scottish Waters

18. *Claymore* 14/19
19. Texaco 15/16
20. *Piper* 15/17
21. Hamilton 15/24
22. *Andrew* 16/28, 16/27
23. *Maureen* 16/29
24. Transworld 21/1, Texaco 20/5
25. *Forties* 21/10, 22/6
26. *Montrose* 22/18, 22/17
27. *Josephine* 30/13
28. *Auk* 30/16
29. *Argyll* 30/24

Norwegian Waters

○1. *Statfjord* 33/9, 33/12
2. Esso 25/8, 25/10, 25/11
3. Bream 17/12

Norwegian Waters

4. Brisling 17/12
5. Flyndre 1/5
○6. *Albuskjell* 1/6, 2/4
○7. *N. Vest Tor* 2/4
○8. *Tor* 2/4, 2/5
○9. *Vest Ekofisk* 2/4
○10. *Ekofisk* 2/4
11. Espen 2/4
○12. *S. Øst Tor* 2/5
○13. *Edda* 2/7
14. *Eldfisk* 2/7, 2/8

Danish Waters

1. *Dan* (S. West Offshore Area)

Dutch Waters

1. NAM F-3
2. Tenneco F-18

○ Fields with significant associated gas reserves.

Table 4.5 (cont)

GAS

Scottish Waters	*English Waters*	*Dutch Waters*

Scottish Waters

1. Total 3/19
2. CFP/Elf/Aquitaine 3/25
3. **BP/IOC 3/29**
4. *Frigg (Scot.) 10/1*
●5. Lomond 23/21
●6. Hamilton 30/2

Norwegian Waters

1. *Odin 30/10*
2. *Frigg (Nor.) 25/1*
3. *N. Øst Frigg 25/1, 30/11*
4. *Elf 25/2*
5. *Øst Frigg*
●6. *Heimdal 25/4*
●7. *Cod 7/11*
8. Murphy 2/3

English Waters

1. Signal 41/24
2. Hamilton 43/20, 43/8, 43/15
3. *Rough 47/3, 47/8*
4. Amethyst 47/14a
5. *West Sole 48/6*
6. Broken Bank 48/18
7. Ranger 48/18b
8. *N. Hewett 48/29*
9. *Hewett 48/29, 48/30, 52/5*
10. Deborah 48/29
11. Dottie 48/30
12. *Ann 49/6*
13. *Viking 49/12, 49/17*
14. Conoco/NCB 49/12, 49/17
15. *Indefatigable 49/18, 49/19, 49/23, 49/24*
16. Mobil/Conoco 49/22
17. Sean 49/25
18. *Leman 49/26, 49/27, 49/28*
19. *Arpet 49/28*
20. Signal 53/4

Dutch Waters

1. *Tenneco K-4*
2. *Petroland K-6, L-7*
3. NAM K-8
4. *Noordwinning K-13*
5. *NAM K-14*
6. *NAM K-15, NAM/Signal L-13*
7. Petroland L-4
8. *Noordwinning L-8*
9. *Placid L-10, L-11*
10. **Zuidwal** (Wadden Zee)
11. Mobil P-6
12. Amoco/Tenneco P-15
13. Sea-Invent Q-7

● Fields with significant associated oil reserves.

Source: Data collected from a wide variety of sources.

TABLE 4.6   NORTH SEA – DISTRIBUTION OF OIL AND GAS DISCOVERIES TO 31 JANUARY, 1975

		Commercial Finds			Others			All Fields		
		Oil	Gas	Total	Oil	Gas	Total	Oil	Gas	Total
Northern*	Scottish Waters	21	2	23	8	4	12	29	6	35
North Sea	Norwegian Waters	9	7	16	5	1	6	14	8	22
	Sub-total	30	9	39	13	5	18	43	14	57
Southern*	English Waters	–	9	9	–	11	11	–	20	20
North Sea	Danish Waters	1	–	1	–	–	–	1	–	1
	Dutch Waters	–	8	8	2	5	7	2	13	15
	Sub-total	1	17	18	2	16	18	3	33	36
	Total	31	26	57	15	21	36	46	47	93

Present Status   Scottish Waters – No fields in production
Norwegian Waters – One oil field (Ekofisk) in production
English Waters – Five gas fields (West Sole, Leman, Indefatigable, Hewett, Viking) in production
Danish Waters – One oil field (Dan) in production
Dutch Waters – No fields in production
German Waters – No discoveries to date

* The line of the Northumbrian Arch and the Dogger Bank High makes for a convenient geological division between the northern and southern North Sea (p. 13), and also coincides with the jurisdiction line between Scottish and English waters (Fig. 3.5).

Source: Data collected from a wide variety of sources.

the first gas ashore from Placid L-10, L-11 in 1975.

In the Danish and West German sectors drilling results have been rather disappointing, although discoveries of traces of gas in several wells in the latter sector in 1974 have raised hopes for the future. In the Danish sector plans are in hand to boost production from the Dan field by the installation of three additional platforms. A number of minor oil and gas strikes have been made, mainly to the north west of Dan, but none are so far of commercial significance. In English waters the main activity in recent years has been the further development and maintenance of production from the five producing fields listed in Table 4.6.

There is a striking contrast in the distributional patterns of oil and gas in the North Sea (Table 4.6). The table illustrates the marked concentration of gas fields in the south while in the north, although oil appears dominant, there are also substantial gas reserves. Estimates of peak production and recoverable reserves for Scottish fields are given in Table 5.2.

REFERENCES AND NOTES

1. *Report on the First United Nations Conference on the Law of the Sea, Geneva, February 24th/April 27th, 1958,* HMSO, Cmnd. 584 (Parliamentary Papers, 1958–59, Vol. 32), London. Particularly relevant are the conclusions of the committee on 'The Continental Shelf'.
2. *Continental Shelf (Jurisdiction) Order 1968,* Statutory Instrument No. 892, HMSO, London.
3. Blocks leased by Shell or jointly by Shell/Esso are explored by Shell UK Exploration and Production Ltd which is commonly abbreviated to Shell Expro.
4. GEORGE, W. J. (1974), 'North Sea—Recent Activity in the UK Sector', *Financial Times* Scandinavia and the North Sea Conference, Oslo, 29/30 April 1974.
5. *Standing Conference on North Sea Oil,* Information Sheet (72) 1, 1 April 1972, North Sea Oil Division, Scottish Economic Planning Department, Edinburgh.
6. *North Sea Oil and Gas: A Report to Parliament* (1973), HMSO, London.

5   Onshore—Impact on the Economy

The ramifications of the offshore industry — the establishment of a legal framework, the complexities of the developing pattern of concessions and licensing procedures, and the steadily emerging map of the Scottish petroleum provinces discussed in Chapter 4 — have introduced a new dimension into the economic geography of Scotland. The purpose of this chapter is to consider the relationship between the offshore developments and the growing onshore industry, and to outline the principal characteristics of this onshore activity. It is first necessary, however, to review the whole spectrum of energy reserves, production and consumption in the Scottish context, to provide a basis for the consideration of the developing economic structure of the onshore industry and its socio-economic implications.

Part I: Reserves, Production and Consumption of Energy in Scotland.

1. Estimates of Reserves and Production

There has been, and remains, much speculation as to the potential reserves and production rates of North Sea oil and gas. Although many promising geological structures are now known to exist there is no assurance that these will contain oil or gas until exploration drilling and appraisal of the results can reveal what actually lies beneath the sea bed. Over the past few years a large number of rich strikes have been made, especially in the East Shetland Basin (Fig. 3.5), which have firmly established the North Sea as a rich oil and gas province. Until the potential of known structures can be assessed more fully on the basis of the current drilling programme, however, and until exploration of remaining areas of the sea bed is complete, the full petroleum potential of the North Sea will remain unresolved.

Forecasts of world reserves and North Sea reserves by the oil industry and Odell have already been discussed (p. 4). Accurate estimates of Scottish reserves in the North Sea, however, cannot be readily obtained. Those which are available range from the unduly cautious to the seemingly wildly optimistic. Official Government estimates of both reserves and rates of production have tended since the beginning to err on the conservative side,[1] while uncertainties about the intentions of successive UK governments concerning the rates of taxation to be levied on North Sea oil seem to have created a situation in which it is in the oil industry's interest to understate reserves. At the other extreme, Professor Odell's

TABLE 5.1   NORTH SEA OIL – PROVEN AND ESTIMATED RECOVERABLE RESERVES OF SCOTTISH FIELDS

	Barrels x 10^9	*Tonnes* x 10^9
Proven recoverable reserves[a]	15.5 – 15.7	2.1 – 2.15
Oil industry estimate of total recoverable reserves[b]	26.0	3.5
Government estimate of total recoverable reserves[c]	22.1	2.9
Odell's estimate of total recoverable reserves[d]	40.0 – 70.0	5.5 – 10.0

a   Based on latest (January 1975) available data obtained from the petroleum industry and *Petroleum Times*. A breakdown of reserves of individual fields is given in Table 5.2.
b   See reference 3.
c   See reference 5.
d   See reference 2.

estimates[2] are certainly too optimistic in the light of current knowledge. His computerised forecasts are presently based on data which does not distinguish between the different national sectors, although he reckons that Scottish reserves make up about one-half of the total North Sea potential. We contend, therefore, that true figures of both proven recoverable reserves and total recoverable reserves are in excess of the oil industry estimates given in Table 5.1 although

by how much is open to conjecture in the absence of available data. However, since we consider the oil industry's figures to be those closest to the true estimates of reserves they have been used widely in this chapter.

On the basis of oil industry estimates and providing production schedules projected in 1974 are realised, Scottish fields could be producing over 3.3 million barrels of oil per day (165 million tonnes/

TABLE 5.2   NORTH SEA OIL—DATA FOR SCOTTISH FIELDS, 31 JANUARY 1975

Field	Block Number	Commencing Year of Production	Estimated Year in which Peak Production Attained	Estimated Peak Production barrels/day $\times 10^3$	Estimated Recoverable Reserves barrels $\times 10^6$
EAST SHETLAND BASIN					
Magnus	211/12	1980	1981	200	1 000
	211/17	?	?	?	?
Thistle	211/18 ⎫ 211/19 ⎭	1976	1978	200–250	800
Dunlin	211/23	1976	1978	220	800
	211/24	?	?	?	?
Cormorant	211/26	1976	1978	150	750
Hutton	211/28 ⎫ 211/27 ⎭	1977	1979	200	1 000
Brent	211/29 ⎫ 3/4 ⎬	1976	1979	450	2 250
	211/24 ⎭	?	?	?	?
Heather	2/5	1977	1979	125	500
Ninian	3/3 ⎫ 3/8 ⎭	1977	1980	400	2 000
Alwyn	3/14	1977	1979	100–150	500–700
Beryl	9/13	1975	1977	100–150	750
Sub-total		—	—	—	10 350–10 600[a]
CENTRAL NORTH SEA BASIN					
Caymore	14/19	1977	1979	200	800
Piper	15/17	1976	1978	265	800
Andrew	16/28 ⎫ 16/27 ⎭	?	?	100	500
Maureen	16/29	1977	1979	125	500
Forties	20/10	1975	1978	400	2 000
	22/6	?	?	?	?
Montrose	22/18 ⎫ 22/17 ⎭	1976	1977	50–100	200
Josephine	30/13	1977	1978	50	100
Auk	30/16	1975	1977	45	100
Argyll	30/24	1975	1975	40	120
Sub-total		—	—	—	5,120[b]
Total		—	—	—	15 470–15 720[c]

? Still to be evaluated
a 1 380 to 1 413 $\times 10^6$ tonnes
b 683 $\times 10^6$ tonnes
c 2 062 to 2 096 $\times 10^6$ tonnes

Source: The Petroleum Industry
Petroleum Times

annum) by 1980 (Table 5.3). Further, if the oil industry's estimate of discoveries yet to be made is valid, production could reach 4 million barrels/day (200 million tonnes/annum) by the mid-1980s.[3] Thereafter, although decline would set in, the industry estimates that a rate in excess of 2 million barrels/ day could be maintained well into the 1990s. By contrast, Odell considers that his much larger estimate of reserves (Table 5.1) could sustain a peak production rate of at least 6 million barrels/day, and possibly as high as 10 million barrels/day by the early 1990s. He estimates that, commencing in 1976, development capacity will grow annually by an increment of 400–600 000 barrels/day (20–30 million tonnes/annum).[4]

The Department of Energy's 'brown book'[5] published in 1974 estimates a production rate of 2–2.8 million barrels of oil per day by 1980. It goes on to suggest that a production rate of between 2 and 3 million barrels/day should be maintained throughout the 1980s. These figures, like the Government estimate of total recoverable reserves given in Table 5.1, refer to the whole of the UK sector. Bearing in mind the distribution of known fields and potentially rich structures, however, the figures correspond in large measure to Scottish waters.

Production rates and profiles such as those projected in Table 5.3 must be treated with caution. 'Slippage' has been significant in the past and further slippage would obviously delay attaining the target levels given. Lack of clear government policy, delays in the completion of rigs and platforms, technical difficulties, inclement weather and cost overruns have combined to

TABLE 5.3   NORTH SEA OIL–ESTIMATED PRODUCTION RATES OF SCOTTISH FIELDS, 1975–80

Field	Barrels/day x 10^3					
	1975	1976	1977	1978	1979	1980
EAST SHETLAND BASIN						
Magnus	–	–	–	–	–	50
Thistle	–	20	110	200	200	200
Dunlin	–	20	120	220	220	200
Cormorant	–	20	100	150	150	135
Hutton	–	–	20	110	200	200
Brent	–	65	150	300	450	450
Heather	–	–	20	100	125	125
Ninian	–	–	60	250	350	400
Alwyn	–	–	10	60	100	100
Beryl	10	60	100	100	90	80
Sub-total	10	185	690	1 490	1 885	1 940
CENTRAL NORTH SEA BASIN						
Claymore	–	–	40	100	200	200
Piper	–	100	200	265	265	265
Andrew	?	?	?	?	?	?
Maureen	–	–	20	100	125	125
Forties	70	250	400	400	400	400
Montrose	–	25	50	50	50	50
Josephine	–	–	25	50	50	50
Auk	20	40	45	45	40	40
Argyll	40	40	40	40	40	40
Sub-total	130	455	820	1 050	1 170	1 170
Others[a]	–	–	–	–	50	200
Total	140	640	1 510	2 540	3 105	3 310

?   Still to be evaluated

a   In the above estimates only the figures given for 'Others' relate to fields not so far shown to be commercial finds (Fig. 4.6).

Source: The Petroleum Industry
Petroleum Times

delay production targets. It was originally intended, for example, that Scottish crude oil production would reach 500 000 barrels/day by 1975. At the time of writing there are doubts about achieving even the much more modest figure of 140 000 barrels.

Faced with likely shortfalls in production the Government announced details in December 1974 of the powers they propose to take to control the rate of oil production from the North Sea. These include no imposition of delays on finds discovered before the end of 1975 and no cuts in production on these fields until at least 1982 or for a period of four years after the start of production, whichever is the later; no cuts in production from any fields discovered after 1975 under existing licences until 150% of the capital invested is recovered; and at a later date a general limiting to 20% of any cuts which may be required. The last mentioned will take into account technological and commercial aspects of individual fields and the needs of the offshore supply industry.[6]

The various estimates of peak oil production all have one thing in common — they are greatly in excess of Scotland's domestic requirements (Fig. 5.1b) even allowing for a modest increase in consumption between now and 1980. Likewise the flow of gas from the Frigg Field through the St. Fergus pipelines is expected to reach 57 million m^3/day by 1977[7](Table 5.4). This is the calorific equivalent of about 7 300 million therms/annum or about thirteen times Scotland's current needs of gas for fuel (Fig. 5.1c). These facts highlight the growing surplus of Scottish energy production over consumption in the near future. The implications of this form the next topic for discussion.

2. Relationships between Production and Trade

The organisation of the oil industry on a world scale is closely governed on the one hand by the economics of production and on the other by those of processing oil and gas and transporting the resulting refined products and petro-chemicals. Different sets of conditions obtain, however, depending on whether the subject is crude oil or natural gas. In the case of oil, the key economic consideration is to locate refineries as close as possible to markets. It is much easier to transport crude oil in bulk over long distances than to move the multifarious separate cargoes of refined products. With gas, transport by pipeline is relatively simple and cheap but to move it by ship it must first be liquefied. Plant to liquefy small quantities of gas can be accommodated on offshore platforms but where large amounts of gas are involved there is no alternative to piping it ashore (see also p. 30). If the gas is to be re-exported the liquefaction plant is likely to be built near the landfall. Further, if new factories are to be built to use the gas in petro-chemical manufacturing, these are likely to be located near the landfall to obviate further expensive pipeline construction. Thus in the case of offshore gas it is the products produced from it which are most likely to be transported long distances rather than the gas itself, a situation opposite to that pertaining to crude oil. The purpose of this section is to investigate the implications of these facts in a Scottish context, as they are likely to modify the developing framework of the petroleum and petrochemical industries.

There are two basic categories of crude oil, namely heavy and light crudes, and further subdivisions

TABLE 5.4   NORTH SEA GAS—DATA FOR SCOTTISH FIELDS, 31 JANUARY 1975

Field	Block Number	Commencing Year of Production	Estimated Year in which Peak Production Attained	Estimated Peak Production m^3/day $\times 10^6$	Estimated Recoverable Reserves $m^3 \times 10^9$
Frigg (Scot)	10/1	1977	?	20[a]	170

?   Still to be decided.

a   700 ft^3/day $\times 10^6$. Gas from the Norwegian section of the Frigg Field will also be piped to St Fergus

b   Total recoverable reserves of Frigg (Scot.) 10/1 and Frigg (Nor.) 25/1 combined are estimated to be 212$m^3 \times 10^9$, that is, 181.7 $\times 10^6$ tonnes oil equivalent. This figure does not include estimated reserves of 100$m^2 \times 10^9$ in other Norweigan fields within the Frigg group (Table 4.5). Some oilfields are also known to contain substantial reserves of associated gas. For example, Brent is capable of producing 17m^3/day $\times 10^6$.

Source: The Petroleum Industry
Petroleum Times

of them according to their content of sulphur and other extraneous minerals. Those crudes which consist predominantly of heavy fractions are the source particularly of fuel oils, for example boiler fuels for industry and ships. Middle East and other imported crudes, notably Nigerian and Iranian, fall into this category. By contrast, a large percentage of the petroleum discovered to date in the North Sea comprises crude oil containing a high proportion of lighter fractions, suitable for the production of diesel oil and motor spirit (petrol). Middle East crudes also tend to be 'sour', that is they contain significant quantities of sulphur which tends to produce undesirable pollutants, while North Sea crudes are virtually sulphur-free ('sweet') making them cheaper to refine.

The implications of these crude oil characteristics for the trade patterns associated with North Sea oil are at present difficult to assess. Indeed, the details of this trade lie beyond the scope of this chapter. The need for fuel oil in Scotland, however, is such that imports of heavy crudes from the Middle East will continue. On the other hand, the quantities of 'sweet' crude oil which will be produced from the North Sea may conceivably meet a large part of West European requirements for diesel oil and motor spirit. At present there is a UK Government stipulation that oil extracted from fields in the UK sector must be landed in the UK. Once companies have fulfilled this stipulation they will be free to ship oil to the market of their choice anywhere in the world. Consequently, there is likely to arise an important Scottish export trade in crude oil.

At present there is only one major refinery in Scotland, the BP plant at Grangemouth, which is designed to supply the specific needs of the Scottish market. In the early 1970s it was indicated that the refinery capacity at Grangemouth might be doubled by 1977.[8] However, plans published in 1972 for the Hound Point terminal and the pipeline to it from the Forties Field showed that even with this increase in capacity Grangemouth would still be able to process only about one-half of the Forties production.[9] With the announcement in November 1974[10] that the refinery expansion had been shelved meantime it became certain that most of the crude petroleum from Scottish fields will be exported (pp. 79 and 101).

There have been multiple proposals for refineries around the Scottish coast in recent years but so far to little effect. Of proposals which have reached the stage of advanced planning, three on the Clyde have been turned down (Fig. 6.16) and one on the Cromarty Firth is soon to be the subject of a public inquiry (p. 79). Further, the

planning of oil terminals in Shetland, Orkney, the Cromarty Firth and Firth of Forth to accommodate super-tankers of 250 000 tons and above, indicates that substantial quantities of offshore oil production may be destined for markets beyond Europe. As it is impossible for fully loaded super-tankers to navigate the relatively shallow waters leading to locations around the southern North Sea, the direction of these exports seems unlikely to be to existing North West European refineries.

The fact that most of the recent proposals for refineries in Scotland have come from US companies suggests that their interest has lain in supplementing the increasingly inadequate US refinery capacity, although such long-distance shipment of refined products is in direct contradiction to the economics of oil refining described earlier. For a number of years no new refineries have been built on the east coast of the US, a situation due almost entirely to the effectiveness of the American environmental lobby. The new emphasis in America on self-sufficiency in energy production,[11] however, seems likely to blunt the opposition to building refineries in the United States, and this could well reduce US interest in refinery projects in Scotland. Consideration of these problems focuses attention upon the details of the Scottish energy consumption pattern.

3. Consumption of Oil and Gas in Scotland

Oil and gas are making a significant and growing contribution to Scotland's energy requirements (Fig. 5.1a). Over the past decade petroleum has replaced solid fuels as the main source of power. It comprised 48% of the total in 1973. Of particular note is the now significant and rapidly growing role of natural gas in meeting energy demands. Also, Scotland had become by 1970 a net exporter of electricity, a situation likely to become more pronounced as the decade proceeds. Between 1964 and 1973 oil consumption more than doubled (Fig. 5.1b) with fuel oil now comprising about 40% of total consumption (which makes necessary continued imports of heavy crudes—see above).

Fig. 5.1c shows that the consumption of manufactured gas, that is supplies derived from coal or light oil distillate, remained fairly steady until 1970. Since then, however, natural gas has made a rapidly growing contribution to total gas supplies (83% of the total in 1973—74). At present natural gas is imported through

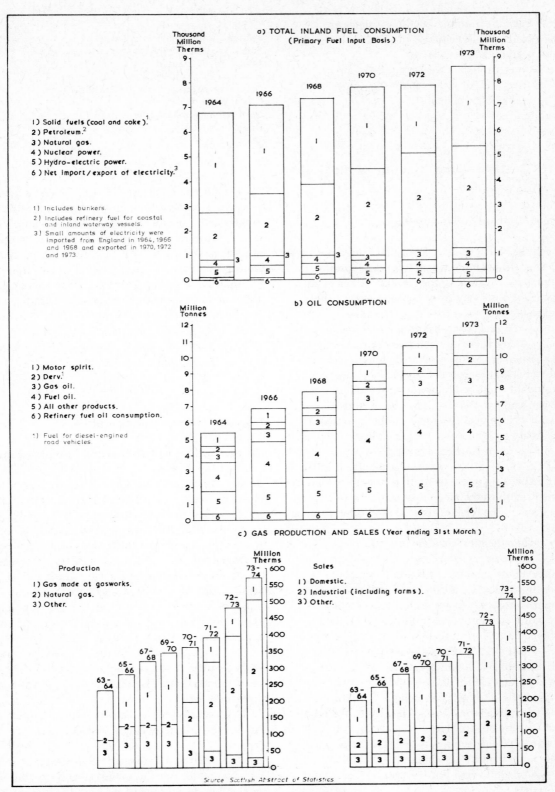

a) TOTAL INLAND FUEL CONSUMPTION (Primary Fuel Input Basis)

Thousand Million Therms

1) Solid fuels (coal and coke).[1]
2) Petroleum.[2]
3) Natural gas.
4) Nuclear power.
5) Hydro-electric power.
6) Net import/export of electricity.[3]

1) Includes bunkers.
2) Includes refinery fuel for coastal and inland waterway vessels.
3) Small amounts of electricity were imported from England in 1964, 1966 and 1968 and exported in 1970, 1972 and 1973.

b) OIL CONSUMPTION

Million Tonnes

1) Motor spirit.
2) Derv.[1]
3) Gas oil.
4) Fuel oil.
5) All other products.
6) Refinery fuel oil consumption.

1) Fuel for diesel-engined road vehicles.

c) GAS PRODUCTION AND SALES (Year ending 31st March)

Million Therms

Production
1) Gas made at gasworks.
2) Natural gas.
3) Other.

Sales
1) Domestic.
2) Industrial (including farms).
3) Other.

Source: Scottish Abstract of Statistics

FIG 5.1   SCOTLAND—ENERGY PRODUCTION AND CONSUMPTION

the grid (Fig. 5.6) from the English gas fields in the southern North Sea. When the Frigg Field is in production this position will be reversed.

	%
Four platforms, equipped and installed	57.45
Submarine lines, complete	14.90
Cruden Bay installation	0.21
Land pipeline, complete	2.98
Oil/gas separation and treatment facilities, Grangemouth	2.98
Forth marine terminal, Hound Point	3.40
Drilling 100 development wells	10.64
Other costs, provisions	7.44
	100

Part II: The Developing Structure of the Onshore Industry.

1. Costs of Developing Offshore Fields

Although the North Sea is a hostile, and therefore a very high-cost environment in which to operate, production costs are low in relation to the current market price of oil. Technical costs vary from field to field but the difference between production costs and the market price of oil remains very substantial in the case of most North Sea fields. It is estimated that even allowing for recent sharp increases in costs (see below) the total capital and operating costs of North Sea production might amount over the life of a normal field to around US $2 per barrel at 1974 prices.[12] Since North Sea crudes are of high quality and low sulphur content they should justify a premium of US S1 per barrel, that is, the value of a barrel of North Sea oil at 1974 prices would be around US $12.[13] Unless world oil prices fall drastically North Sea oil is therefore likely to prove a very lucrative investment both to the oil companies, and also to the governments in the form of royalties and other forms of taxation.

While the great increase in world crude oil prices since 1973 has improved the apparent profitability of North Sea oil development, it is also the major factor contributing to world-wide inflation and, therefore, to development costs in what was already a very high-cost area by world standards. For instance, the original cost of developing the Forties Field was put at £350 million by BP;[14] later, in 1973, this estimate was revised to £450 million (including £50 million interest on capital borrowed up to 1976).[15] In the autumn of 1974 this figure had been raised to £630 million,[16] an increase of 40% in little over a year. Further, it is estimated that a company setting out to develop a field similar to Forties today would have to be prepared to face the prospect of investing more than £1 200 million.[17]

A breakdown in Forties costs on a percentage basis is roughly as follows:[18]

In 1973, the last full year for which figures are available, it is estimated that exploration for oil and gas in the North Sea cost the industry £550 million.[19]

Investment is likely to increase substantially over the next decade as offshore activities expand and the industry moves increasingly into the production phase. Expenditure on such a scale has already encouraged massive complementary developments onshore. These may be broadly classified as follows:

    i Servicing and transport facilities
   ii Engineering for offshore developments
  iii Pipelines, processing and storage.

2. Servicing and Transport Facilities

The pace with which offshore drilling is proceeding has created a rapidly developing need for the provision of services on adjacent coasts. Service bases provide the 'back-up' facilities—materials, transport, catering and other services—necessary to keep both rigs and supply vessels operational. During exploration drilling, a single rig requires on average 1 000 tonnes of supplies per month, largely fuel, drilling chemicals and steel pipes. The production drilling phase will call for an even larger volume of supplies, a single platform using 30 000 tonnes of supplies per annum, including approximately 50 per cent liquid (water and fuel), 25 per cent powder (cement, etc.) and 15 per cent tubular steel.[20]

To support this activity a shore base must be selected with care so as to minimise delays to high-cost offshore operations. The main requirements of a marine service base are:

    i An all-weather harbour, with access at all states of the tide, providing a 24-hour, seven-day service.

    ii Berths with at least 5.5 m (18 ft) depths at low tides and lengths varying between 37 m and 76 m. Preferably berths should be deeper to handle the new larger types of vessels (see below), also cargo vessels and coastal tankers with incoming supplies. Quays must be constructed to accommodate heavy and lengthy loads such as cement, tubular steel, rig anchors and chains.

      iii Storage tanks for fuel and drinking water for both supply vessels and rigs; also silos for drilling cement and muds.

      iv Readily adjacent warehouse and open storage space.

      v Good road, and preferably also rail, access to facilitate handling of supplies.

      vi A centre of population to provide labour, services (including transport) and supplies.

      Helicopter landing facilities may be an additional attraction, but these need not actually be at the base concerned. To limit the time required to turn round a service ship, and thus the number of ships required to service a rig or platform, particular attention has to be paid in all new construction of marine service bases to provide what is termed 'one-stop shopping'. In such a system all of a supply vessel's needs are catered for at the same berth, with fluids and powders being piped into the vessel simultaneously with the crane-

loading of its deck cargo. Pipes in ducts below the quay deck connect the vessel directly to supply tanks and silos at the base.

      But while quick turn-round of vessels can be an important factor in an operator's choice of service base, the overall location pattern of such bases can best be understood by reference to the operating range of service vessels and to the logistics of supplying the bases themselves. Fig. 5.2 shows the sea areas within 320 km (200 ml) of a series of existing and possible service bases, a distance which represents a passage time of between 12 and 14 hours for service vessels. Given the common complement of two general service vessels per rig, and allowing for the time spent loading and unloading in port, operators are not likely to want to depend for their main supplies on bases much more distant than this. Further, this distance also coincides with the maximum operational range of the Sikorsky helicopters which have been most widely used in servicing rigs.[21] The pattern of service vessel operation is steadily becoming more specialised and elaborate, but these conditions were particularly significant a few years ago when most exploration was being done in the Central North Sea Basin. They did much to establish the importance of Aberdeen and Peterhead in

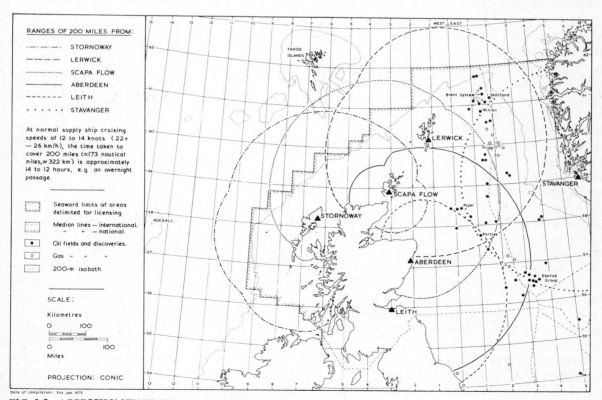

FIG. 5.2. ACCESSIBILITY FROM KEY PORT LOCATIONS

the offshore busines (pp. 84, 91 and 92).

Fig. 5.2 also indicates the definite advantages to be had in servicing the newer fields east and north of Shetland from ports in the islands, also of handling operations in the eastern Atlantic from either the Northern or Western Isles. In practice, however, supply-carrying vessels working in northern waters have tended to use Shetland or Orkney ports mainly as 'forward' bases, hauling their main loads of basic stores from ports such as Aberdeen, Peterhead or Dundee, and making a smaller number of intermediate runs to island bases. A major reason for this is the considerable extra cost of shipping supplies to island bases in the first place, especially the additional expense incurred in double handling of goods delivered by sea.

It should be noted, however, that service vessels are not all of one type. The paragraph above applies particularly to supply-carrying ships, but most service vessels are now multi-purpose ships and a whole variety of further specialist classes of vessel has been designed and brought into service in the past few years. The key elements in the recent sequence of development of service vessels have been:

i A steady increase in the size and power of vessels to enable them to cope with the rougher weather and longer hauls in the northern North Sea. Ships of around 1 000–2 000 bhp and 30–37 m in length sufficed to carry supplies to jack-up rigs of the US Gulf Coast and in the southern North Sea. The equivalent vessels in use today off Scotland are 52–61m long and of 3 000–4 000 bhp.

ii The provision of increasingly powerful winches, derricks and stern rollers to enable vessels to act as tugs and to be able to lift and relay the large anchors which hold semi-submersible rigs and lay barges on station. The larger vessels of this type now in service are up to 67 m in length and develop around 7 500 bhp, using up to six engines to drive their propellers, bow thrusters, winches and generators. In addition to deck space for up to 750 tonnes of cargo they have specialised tank storage for water, fuel, drilling supplies, wire hawsers and anchor chains.

iii The introduction of specialised pipe-carrying ships to haul pipe from ports to lay barges. These can typically carry 140–200 12m (40ft) pipe-lengths, in comparison with the limit of about 40 on general purpose service vessels used for this purpose.

iv The use of some of these pipe carriers, or of other ships specially designed to carry large consignments of materials, to meet the massive requirements of production platforms when these are at the well drilling stage.

v Additionally, a number of large sea-going tugs and salvage vessels are now on constant stand-by to service the rigs, and a considerable fleet of unmanned barges is being used to carry modules and other steel structures out to the oil and gas fields under development.

Because of the particular jobs they do the various craft which serve the offshore industry make differing demands on service base facilities. Pipe carriers obviously operate mainly out of a limited number of ports, such as Leith and Invergordon where pipe coating is undertaken, or Peterhead where pipes are stockpiled. Pipelaying is highly dependent on spells of good weather during which the lay barges must have vast quantities of materials shifted to them over a short period. Peterhead, with its proximity to the working areas, offers special advantages in this, and Shetland bases will become increasingly important over the next few years. On the other hand, such ports can suffer sharp drops in trade in the winter season or when particular pipelaying contracts are completed. Island bases can readily provide the bunkers and ships' stores which are the main requirement of service ships or tugs which specialise in anchor handling and the towing of rigs, but platform supply vessels are likely to use mainland bases because of the vast quantities of stores which they are required to handle. With the considerable increase in the size of service vessels, quays which a few years ago were capable of handling four vessels at a time may now be classed as 3-berth units, and so on.

Fig. 5.3 shows the service bases developed to date: a major grouping on the east coast, a secondary one in Shetland, and one small facility in Orkney which is used on a temporary basis only. Figures for oil-related movements are not available in the same form for all of these ports, and they may refer to quite differing types of vessel. Aberdeen, for instance, is concerned almost entirely with service ships and survey vessels which bring considerable trade to service bases. Peterhead's traffic includes barges and vessels sheltering from bad weather, which may make quite limited demands on commercial facilities. However, on the basis of data collected for the period July–December 1974 we reckon that the division of trade is approximately as follows: Northern Isles—11% (almost all at Lerwick); Peterhead—46%; Aberdeen—31%; Dundee, Leith and Montrose combined—12%. With an increasing number of bases coming into operation, and with pipe-laying and production drilling activities about to expand in the East Shetland Basin it can be expected that the Northern Isles' share of the total traffic will rise over the next few years.

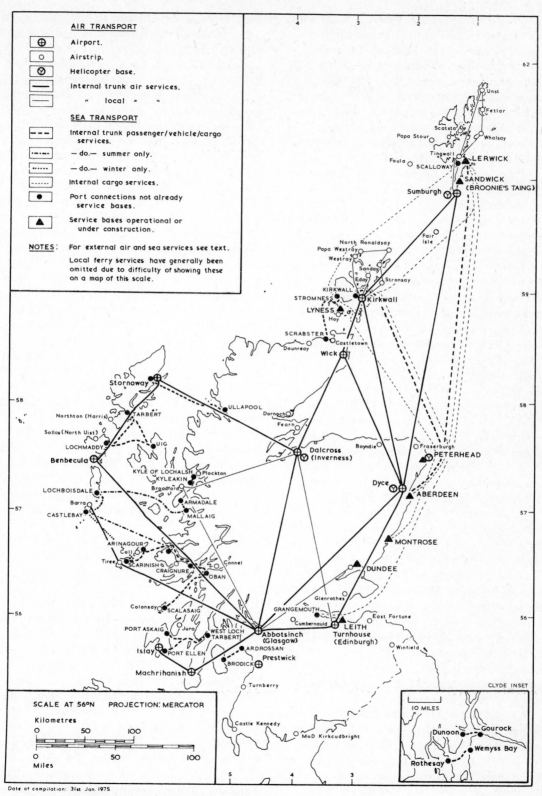

FIG 5.3   INTERNAL AIR AND SEA ROUTES

Date of compilation: 31st Jan. 1975

Establishment of the oil industry has led to a remarkable expansion in both internal and external air services, and has also supported the establishment of a significant number of new sea freight services. Fig. 5.3 shows the main internal routes. Quite apart from the development of major helicopter bases at Aberdeen and Sumburgh which ferry crews and urgent stores to and from the rigs, oilmen have shown themselves to be extremely ready to pay for the speed and convenience of air travel. Expansion of services has variously taken the form of much increased flight frequencies on existing routes, for example Aberdeen—London and Aberdeen—Shetland, and the establishment of wholly new routes, mainly operated by companies new to the area. The most spectacular traffic growth has been at Aberdeen (p. 90) and Sumburgh, both of which handle a great many charter and air taxi flights as well as scheduled services. External services from Aberdeen now include direct daily flights to Stavanger and Amsterdam and a special link service to Prestwick connecting with direct flights to Houston—all major centres for the oil industry.

New sea freight routes include links between Boston (Lincs.), Grangemouth, and the Northern Isles; between the Humber, Grangemouth, Lerwick and Stavanger; also services from Aberdeen to Orkney and Shetland which parallel the established North of Scotland Shipping Company routes, and links between Aberdeen and the Netherlands. There has also been a considerable development of unscheduled vessels carrying oil-related cargo to all the main service bases and a heavy traffic in construction materials to platform yards on the west coast. Additionally, cargo liners from US Gulf ports now call at least twice monthly at Aberdeen, and also at Leith and Dundee, with part loads of oilfield machinery and stores.

On land, roads handle the bulk of oil-related traffic. Locally this has given rise to the spectacularly rapid expansion of some road haulage fleets and has put damaging pressures on roads in the neighbourhood of major construction schemes (p. 72). In some cases local authorities, before giving planning permission, have insisted that certain developments be treated as 'island' sites (p. 107). In others development has necessitated the upgrading of short stretches of existing roads or the construction of new access roads.

A major problem is, however, that on a national basis roads in what are now key development areas, such as the North East and round the Cromarty Firth, have received a lower priority than those in the Central Lowlands. The inadequacy of the national trunk road network is well illustrated by Fig. 5.4.

With the exception of about 10 km of dual carriageway between Aberdeen and Stonehaven there are no other lengthy stretches of dual carriageway north of Dundee, while motorways are almost entirely restricted to the Central Belt. Following considerable pressure from local authorities and others, the go-ahead for the reconstruction of the A9 between Perth and Inverness was given in 1974, but only a few short stretches will be built as dual carriageway and those only where safety demands. North of Inverness the A9 will be completely realigned between Inverness and Invergordon (p. 80).

North of the Tay the carrying capacity of the rail network is severely limited by the fact that lines are single track over long stretches with few passing places. Nevertheless, a new interest in the use of the railways for both passenger and freight traffic has been generated by oil-related activities in the North and North East. Oil-related developments in the Loch Carron area may well give a new lease of life to the Kyle line (Dingwall—Kyle of Lochalsh). Already Alness station has been reopened to passenger traffic (p. 80) and a Motorail service introduced between Stirling and Inverness has proved attractive because of the difficulties of negotiating the A9. The railways have played a significant part in transporting steel tube for both land and submarine pipelines. A new spur has been built to serve M—K Shand's yard at Saltburn (p. 80) and Waterloo Goods Yard in Aberdeen has been reconstructed as a well-casing depot (p. 86). Additionally, pressure has been exerted to have the branch line to Peterhead relaid to serve the needs of the new industrial complexes in the area already under construction or planned.

3. Engineering for Offshore Developments

This group embraces all industrial activities connected with the manufacture of equipment and materials directly required in the exploration for and production of crude oil and natural gas. It covers the manufacture of equipment to transport offshore oil and gas but not that of refinery or petrochemical plant. In terms of location its activities can be broadly distinguished as belonging to one of three categories:

i Industrial activities which can be located only at a limited range of coastal sites with deep water close inshore—specifically production platform construction.

ii Activities which must also be coastal but which are attracted to established areas of shipbuilding

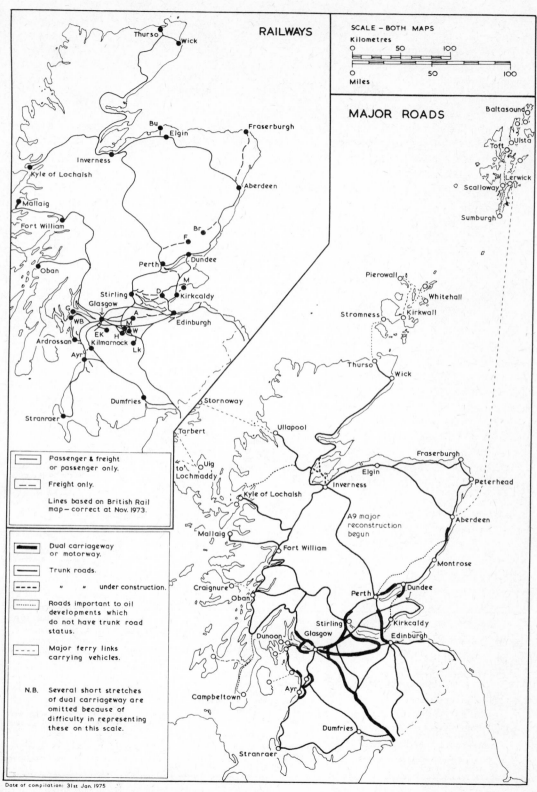

RAILWAYS

SCALE – BOTH MAPS

MAJOR ROADS

Passenger & freight
or passenger only.

Freight only.

Lines based on British Rail
map – correct at Nov. 1973.

Dual carriageway
or motorway.

Trunk roads.

"   "   under construction.

Roads important to oil
developments which
do not have trunk road
status.

Major ferry links
carrying vehicles.

N.B. Several short stretches
of dual carriageway are
omitted because of
difficulty in representing
these on this scale.

Date of compilation: 31st Jan. 1975

FIG 5.4   MAJOR ROADS AND RAILWAYS

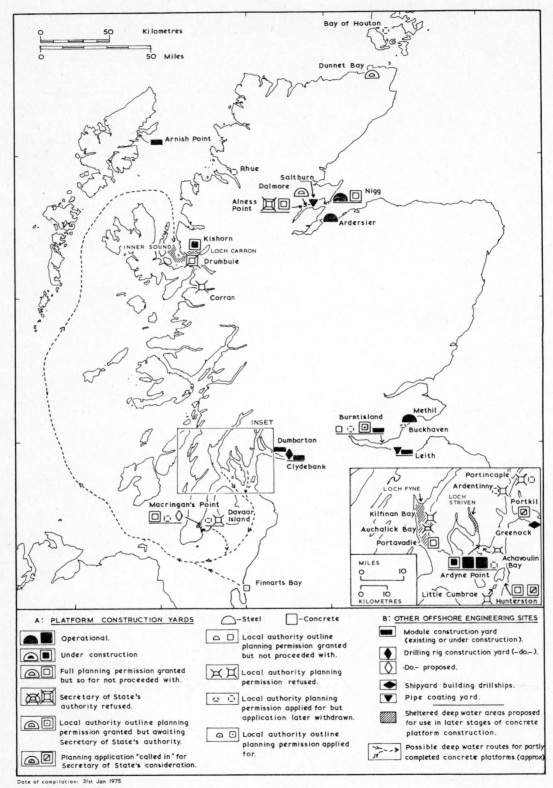

Scale:
0 — 50 Kilometres
0 — 50 Miles

Bay of Houton

Dunnet Bay

Arnish Point

Rhue

Saltburn
Dalmore
Alness Point
Nigg
Ardersier

Kishorn
LOCH CARRON
Drumbuie
INNER SOUND

Corran

INSET

Dumbarton
Clydebank

Burntisland
Methil
Buckhaven
Leith

Macringan's Point
Davaar Island

Finnarts Bay

Inset (Loch Fyne / Clyde):

Portincaple
Ardentinny
LOCH STRIVEN
LOCH FYNE
Kilfinan Bay
Portkil
Auchalick Bay
Portavadie
Greenock
Achavoulin Bay
Ardyne Point
Little Cumbrae
Hunterston

MILES
0 — 10
0 — 10
KILOMETRES

A: PLATFORM CONSTRUCTION YARDS ⌒ —Steel ▢ —Concrete **B: OTHER OFFSHORE ENGINEERING SITES**

◖ ■	Operational.
◖ ▣	Under construction.
◠ ▢	Full planning permission granted but so far not proceeded with.
⊠	Secretary of State's authority refused.
◖ ▣	Local authority outline planning permission granted but awaiting Secretary of State's authority.
◖ ▨	Planning application "called in" for Secretary of State's consideration.
◠ ▢	Local authority outline planning permission granted but not proceeded with.
⊠	Local authority planning permission refused.
◠ ▢	Local authority planning permission applied for but application later withdrawn.
◠ ▢	Local authority outline planning permission applied for.

▬	Module construction yard (existing or under construction).
◆	Drilling rig construction yard (–do–).
◇	–Do.– proposed.
◢	Shipyard building drillships.
▼	Pipe coating yard.
▨	Sheltered deep water areas proposed for use in later stages of concrete platform construction.
⇄	Possible deep water routes for partly completed concrete platforms (approx)

Date of compilation: 31st Jan 1975

FIG 5.5   PLATFORM, RIG AND PIPE COATING YARDS

54

and heavy engineering—construction of drilling rigs, drill-ships and platform modules.

iii Activities which can become established in any area which has a suitable supply of skilled labour and experienced management or which can readily attract such people to it.

The first and second categories are illustrated in Fig. 5.5, but the third could only be shown on this scale by a generalised representation of the main manufacturing areas of the country, and even this would have omitted a scatter of firms in other areas.[22] The pipe coating plants at Saltburn and Leith do not fit clearly into any of the categories but will be considered with the second one.

Construction of production platforms has particularly limiting site characteristics because of the depths of water required and the areas of land needed alongside the deep water. These requirements differ quite

specifically between steel and concrete designs, and at different stages in the construction of concrete platforms, so it is important to distinguish carefully between them. (Table 5.5 lists platform orders placed to date, together with their construction sites.)

The first point to be made is that what are popularly called steel platform yards usually build only the jacket, or leg structure of the platform (see Glossary). This is typically built on its side, to be floated out to its intended site on a barge and then upended into position on the sea bed. Jackets may be built either in dry docks on top of the barge which is to transport them, or in yards from which each completed jacket will be winched out on to a barge brought directly alongside. The superstructure of steel platforms, containing the drilling machinery, power plant, living quarters and so on, are normally built separately in other yards as a series of large

TABLE 5.5 NORTH SEA OIL—PLATFORM ORDERS FOR SCOTTISH FIELDS, 31 JANUARY, 1975

Field	Platform Type	Contractor	Yard	Date of Installation
Magnus	–	–	–	–
Thistle	Steel	Laing Pipelines Offshore	Graythorp, Teesside	1976
Dunlin	Concrete	Andoc	Rotterdam initially	1976
			*Inner Sound finishing**	
Cormorant	*Concrete*	McAlpine/Seatank	*Ardyne Point initially*	
			Inner Sound finishing	1976
Hutton	–	–	–	–
Brent	*Steel*	*Redpath Dorman Long*	*Methil*	1975
	Concrete	Høyer-Ellefsen/Aker/Selmer	Stavanger	1975
	Concrete	*McAlpine/Seatank*	*Ardyne Point*	1975
	Concrete	Høyer-Ellefsen/Aker/Selmer	Stavanger	1976
Heather	*Steel*	*McDermott*	*Ardersier*	1977
Ninian	*Concrete*	*Howard Doris*	*Kishorn*	1977
	Steel	*Highlands Fabricators*	*Nigg Bay*	1977
Alwyn	Concrete	Howard Doris	Stramstad, Sweden	1977
Frigg	Concrete	Howard Doris	Andalsnes, Norway	1975
	Concrete	McAlpine/Seatank	*Ardyne Point*	1975
	Steel	UIE	Cherbourg	Under Installation
Beryl	Concrete	Hoyer-Ellefsen/Aker/Selmer	Stavanger	1975
Claymore	Steel	UIE	Le Havre	1976
Piper	Steel	UIE (lower part)	Le Havre }	1975
		McDermott (upper part)	*Ardersier* }	
Andrew	–	–	–	–
Maureen	–	–	–	–
Forties	Steel	Laing Pipelines Offshore	Graythorp	Installed
	Steel	*Highlands Fabricators*	*Nigg Bay*	Installed
	Steel	Laing Pipelines Offshore	Graythorp	1975
	Steel	*Highlands Fabricators*	*Nigg Bay*	1975
Montrose	Steel	UIE	Cherbourg	1976
Josephine	–	–	–	–
Auk	*Steel*	*Redpath Dorman Long*	*Methil*	Installed
Argyll	Production using the modified semi-submersible Transworld 58 as a temporary platform			1975

Platforms already built or under construction in Scotland are shown in italics.
*Awaiting a final decision.

modules which are transported independently to the operating site and lifted into position once the jacket has been fixed to the sea bed.

In contrast, concrete platforms are built upright. They normally consist of a massive base, within which oil can be stored and from which thick concrete pillars rise to hold the deck clear of the surface of the sea (Fig. 3.7). This deck on which the modules stand is variously of steel or concrete construction. The first stage of building the base takes place in a dry dock, but it must then be floated directly into deep water into which it sinks progressively as construction of the base and pillars proceeds. Completion of the platform is then achieved by ballasting the structure to make it sink completely into the sea. The deck is floated into position on top of the pillars and the modules installed. By deballasting, the platform is then allowed to float high in the water for towing to its operational site. It arrives at the site virtually complete and is lowered to rest on the sea bed, where it maintains its position wholly by reason of its ballasted mass. (See also p. 18).

Construction of concrete platforms requires smaller land areas than are needed for steel jackets, but very much greater water depths with more assured shelter from wind and waves. As little as 7 ha will suffice for the dry dock and associated plant compared with the 40–70 ha which are required to accommodate the building site, sub-assembly yards and fabrication sheds of a steel jacket yard. But whereas jackets can be floated out on barges drawing less than 9 m of water, even the intermediate construction phases of concrete platforms need depths of 45–90 metres and the final submerged testing and deck assembly requires up to 215 m. Because concrete platforms must float during much of their construction, thoroughly sheltered conditions are essential. Further, with the completed units drawing as much as 40 m, the choice of route along which they can be towed to their final destination is very restricted (Fig. 5.5 and p. 105).

Sites of the type required for concrete platforms are most commonly found on the west coast, are liable to be remote, to lack an industrial infrastructure and to be located in areas of high landscape value. It is largely for these reasons that controversy over the establishment of concrete platform sites has been so much greater than that over steel platform yards. Sites with the extreme depths suitable for the completion of concrete platforms have been distinguished in only the Inner Sound/Loch Carron area and in Loch Fyne (Fig. 5.5), but it is not yet clear whether platforms completed in the latter could be safely floated through the channels of the Firth of Clyde and those leading from it to the open sea (p. 105). It

may be that all concrete units built further south will have to be towed to the Inner Sound for completion.[23] As it is, all the yards operating or proposed in the Firth of Clyde will be producing platforms specially designed or modified to cope with construction and navigation depths less than those of the Norwegian-built platforms so far favoured by the oil industry (p. 105). East coast sites proposed for concrete platforms being much shallower still will require yet more highly innovative designs. It may well be significant that none of the yards sanctioned or proposed on the east coast has so far landed an order.

Depths in the Cromarty Firth, Inner Moray Firth and Firth of Forth have proved fully adequate for steel platform building, and are associated with flat coastal areas large enough for the yards and/or drydocks involved. These areas also offered transport facilities which could handle the large quantities of materials needed by the yards, and they had the advantage of being relatively close to the fields where the platforms were to be installed. This last point is important, because it must be remembered that when oil companies and contractors began, only a few years ago, to construct platforms large enough for the depths involved in the central North Sea, they were embarking on projects beyond the existing limits of technology. Much of the detail of how to build, transport and locate the platforms had to be worked out while the projects were actually in progress. It made sense, therefore, to avoid further complications by ensuring the shortest possible sea journeys for the completed structures. In a similar fashion, the first sites for concrete platform construction were in the nearest suitably deep water of the Norwegian fjords, and it is only more recently that sites on the west coast of Scotland have been considered.

The second group of offshore engineering concerns involved in the offshore oil industry—constructing drilling rigs, drill-ships and platform modules— obviously also need waterfront sites, but are much less limited than platform builders in their choice of location. Large as some of these structures may be, none requires depths greater than 10 m for launching and some need as little as 2–3 m. What really counts in determining where to build them is the availability of skilled labour and experienced management. The work force for module construction can be drawn from either shipbuilding or heavy engineering, and in some cases former shipyard sites are being re-used by this new industry (pp. 99 and 104). Drill-ship construction is specifically the sphere of experienced shipbuilders, and all the offshore drilling rigs which have been built in the UK have been produced at existing or former shipyards. In contrast, pipeline coating plants largely make use of semi-skilled labour which can be recruited in most areas and trained over a short period. They may both receive raw materials and send out coated pipe by sea transport (p. 100), but as they also bring in much pipe by rail, and coat pipe for overland as well as undersea schemes, there is no absolute requirement that they should be located at ports.

As modules and coated pipe can be readily shipped to any North Sea field from any yard on the West European seaboard, Scottish manufacturers have to compete with a large range of foreign contractors, in addition to yards elsewhere in the UK. Orders for the modules of a platform are typically placed with firms in two or three different countries. For drill-ships and rigs the market and the competition are effectively world-wide. Drilling units can be towed or moved under their own power to any offshore area where they are required. It is particularly significant in this respect that the only rig-building yard in Scotland (p. 104) is an offshoot of a well-established US firm, and that no real attempt has been made anywhere in the UK to match the production of large semi-submersible rigs in Norway, Sweden and West Germany. There is real hope for the future, however, in the way that the Scott-Lithgow group at Greenock is becoming established in drill-ship construction. Extension of offshore drilling into ever deeper waters could give rise to an increasing and fairly long-term demand for such craft, and advantage will clearly lie with those firms which have an established expertise in their construction.

The third, largely 'foot-loose', group of engineering industries similarly offers the potential of a world-wide market and long-term return to firms which choose to diversify in this direction. In part it is concerned with the production of specialised drilling and underwater equipment for which the oil industry has established US and European suppliers. To develop in these directions expertise may have to be bought by hiring experienced personnel or by entering into licensing arrangements with foreign firms. Over a very wide range of products required by the offshore industry, however, basic engineering skills and existing product lines are directly marketable, given sufficient interest and determination on the part of their producers.

Development of this potential is a prime concern of the Offshore Supplies Office (Fig. 7.1) which has its headquarters in Glasgow thus making it readily accessible to Scottish industry. But success in becoming established as a supplier to the oil industry depends primarily on the firms themselves. The oil industry is a brash, volatile business which handles astronomical sums of money and

works to extremely tight schedules. It expects its suppliers to advertise their wares actively, to turn out accurately machined reliable products, and at all costs to deliver on time. Firms which are prepared to meet these requirements can expect a rapid and rewarding increase in business. With the offshore search for oil and gas now being world-wide, the potential market for Scottish engineering firms could be very significant.

4. Pipelines, Processing and Storage

The pipelines and onshore installations so far developed or projected for Scotland are shown on Fig. 5.6; also the oil pipelines from Finnart to Grangemouth and the pipe which brings gas from the southern North Sea to Central Scotland. Refinery developments have already been considered (p. 46) as well as the selection of routes for underwater pipelines (pp. 31), but fuller attention to the choice of pipeline landfalls and overland routes is appropriate here.

Selection of landfalls has been influenced by two main factors:

i The general need to make underwater pipelines as short as possible because they are so much more expensive to construct than routes overland (eg. over £500 000 per km compared with about £80 000 per km respectively, for the Forties pipelines).

ii Whether the landed product is to be transported onward by ship or by a further pipeline.

The difficulties indicated above (p. 45) in transporting gas by sea make a mainland landfall all but mandatory for gas pipelines (eg. the oil from Brent is to be piped to Shetland but the gas will probably go to St. Fergus). Where oil is to be tanker-loaded for its further journey, however, termination of the pipeline at anchorages suitable for loading ships takes precedence over other factors in deciding the landfall (eg. oil from Piper is to go to Orkney rather than by a shorter pipeline to Caithness or North East Scotland).

Once the general area of the landfall has been determined its specific site is likely to be a compromise between the oil company's search for a site physically favourable to the operation and pressures exerted by various public interests. Ideally, pipelines should be brought ashore across sandy beaches backed by gentle slopes, but in the Northern Isles they may have to be laid in a trench blasted out of solid rock (p. 23). Construction of the actual landfalls can usually be done with extremely little disturbance to the environment, but

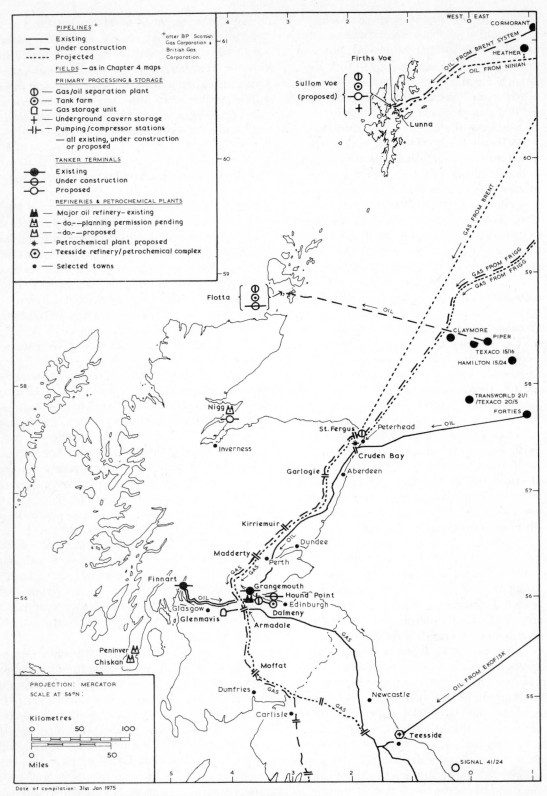

FIG 5.6   PIPELINES, PROCESSING AND STORAGE PLANTS

compressor stations or processing plants associated with landfalls can be much more open to objections (p. 91).

Overland, both oil and gas pipelines can negotiate quite steep gradients, and in general take fairly direct routes from the landfall to their destinations. They are, however, most carefully aligned to avoid their passing through settlements or through woodland, to make them cross rivers, major roads and railways as directly as possible, and to provide for the installation on gas lines of compressor stations at regular intervals (p. 91 and Fig. 5.6). Functionally, the most notable feature of the pattern of pipelines and tanker terminals now becoming established in Scotland is the way in which it is designed to facilitate the export of oil and gas to England and abroad (p. 46).

Storage for crude oil is being provided in substantial tank farms at shipment points such as Hound Point, Flotta and Sullom Voe (also underground storage at the latter), and plant installed for removing associated gases from crude oil at Grangemouth, Flotta and Sullom. Application for planning permission has been made for a petro-chemical plant using natural gas at Peterhead, and others may follow it (p. 91). The plant at Glenmavis near Airdrie which currently liquefies natural gas from the southern North Sea, and can store 40 000 tons of LNG, will perform the same function for Frigg Gas. It provides for peak demand periods and stores sufficient liquid to cover the total gas needs of Scotland for four days.

Part III: Socio-Economic Implications

The exploration, production and refining of oil have been accompanied in most parts of the world by spectacular and far-reaching socio-economic changes. The rapid growth of the oil industry, in addition to creating many new opportunities in Scotland, has put a considerable strain on the labour resources and infrastructure of certain areas. Some of the problems arising are examined in this section, prefaced by a survey of the business structure of the industry.

1. Organisation of the Oil Industry

At first sight the oil industry appears to be immensely complex. However, on closer examination it is possible to rationalise this complexity by reference to the principal operations necessary in the production of crude oil and gas. Fig. 5.7 shows in summary the sequence in time of the operations involved in finding and developing offshore resources; also the relationships between them and the main services they require. All the various phases, except platform operation, are currently represented in Scotland, and the full range of services. Platforms are expected to be operating in late 1975.

The business structure behind this functional organisation is important because on it depend the ways in which decisions are being taken, both within the industry and relative to the world outside. It will be noted that the matrix of Fig. 5.7 does not include the transportation and refining or processing of oil or gas. These operations, and the distribution of the manufactured products, are undertaken wholly by the various oil companies and by the nationalised gas corporations. Because they are commonly in the public eye it is often thought that the same companies and corporations also undertake the earlier operations of finding and producing the oil and gas, but in practice this is not so. Although exploration and production are financed and controlled mainly by the international oil companies, most of the actual work is delegated to specialist firms within the industry, while certain services and ancillary industries are run by large corporations which are not directly connected with the oil industry.

Thus, with the notable exceptions of BP and Shell Expro which operate on their own behalf, geophysical exploration and exploratory drilling are normally undertaken for oil companies by specialist firms which own and operate the necessary vessels and laboratories. Downhole engineering services such as well logging, directional drilling or the flow-testing of oil or gas discoveries are the preserve of yet other companies with the necessary equipment and expertise. Similarly, the building and installation of platforms and pipelines are carried out by various civil engineering consortia, and the complex of service facilities is operated by specialist service companies, harbour authorities and individual contractors. Some supply ship companies are concerned wholly with this line of business, while others are subsidiary units within large shipping combines and others again are linked to service base companies or firms running ocean-going tugs. Catering and engineering requirements may be met both by existing local

FIG 5.7 OIL-RELATED OPERATIONS AND REQUIREMENTS

Key:
- ● required
- ○ not required

SERVICES AND FACILITIES (Individual Categories) → OPERATIONS arranged in sequence of development ↓	Accommodation and Catering — Housing	Work camps and other temporary accommodation	Contract catering, laundry services, etc.	Ports — All-weather harbours	Very deep water close inshore	Transport — Service vessels and tugs	Cargo vessels and barges	Rail transport	Heavy road haulage	Scheduled and charter air services	Helicopter services	Communications — Standby safety vessels	Navigational aids	Telecommunications	Supply Handling and Storage — Open storage	Warehousing	Specialised storage (silos, tanks, etc.)	Customs documentation, Freight handling, etc.	Engineering — Dredging	Land reclamation	Machinery repairs and servicing	Steel fabrication	Plant hire (compressors, welding and drilling equipment, cranes, etc.)	Industrial gases	Muds and mineral fluids	Diving services	Specialised drilling services	Engineering and scientific consultancy
1. Geophysical and Oceanographic Exploration	○	○	○	●	○	○	○	○	○	●	○	○	●	●	○	●	●	●	○	○	●	○	○	●	○	○	●	●
2. Construction of Service Bases and Ports	○	●	●	○	○	○	●	●	●	●	○	○	○	○	●	○	●	●	●	●	●	●	●	●	○	○	○	●
3. Oil Rig Operation	●	○	●	●	○	●	○	○	●	●	●	●	●	●	●	●	●	●	○	○	●	○	●	●	●	●	●	●
4. Pipe-Laying (Sea)	○	●	●	●	○	●	●	●	●	●	●	●	●	●	●	●	●	●	●	○	●	●	●	●	○	●	○	●
4. Pipe-Laying (Land)	○	○	●	○	○	○	○	○	●	●	○	○	○	○	●	○	●	●	○	○	●	○	●	●	○	○	○	●
5a. Production Platform Construction	●	●	●	●	●	●	●	●	●	●	●	○	●	○	●	●	●	●	●	●	●	●	●	●	○	●	○	●
5b. Module Construction	●	●	●	●	○	○	○	○	●	●	○	○	○	○	○	○	●	●	○	○	●	○	●	●	○	○	○	●
6. Production Platform Installation	●	●	●	●	○	●	●	○	○	●	●	●	●	●	●	●	●	●	○	○	●	●	●	●	●	●	○	●
7. Production Platform Operation	●	○	●	●	○	●	○	●	○	●	●	●	●	●	●	●	●	●	○	○	●	●	●	●	○	○	○	●
Management of Exploration and Production	●	○	○	○	○	○	○	●	○	●	○	○	○	●	○	○	●	●	○	○	○	○	○	○	○	○	●	●

Note: Other existing local services which have to be greatly expanded to serve the oil industry include:

Hotels, licenced premises, restaurants, car hire and taxi services, travel agencies, banking and insurance, secretarial services, property agencies (lawyers and estate agents), employment agencies

Building and construction trades are particularly heavily employed, with labour shortages quickly developing enterprises or by companies specially set up for the purpose.

Although the most specialist engineering aspects of the oil industry may remain to a considerable extent outwith the scope of Scottish-based enterprise for the immediate future,[24] there is a developing pattern of Scottish business interests within the industry. Of particular importance is the part played by Scottish institutions in raising finance. The Scottish clearing banks have set up specialist offshore departments and are engaged in strengthening international financial links to cope with large-scale developments.[25] These banks, together with the Scottish-based merchant banks, investment trusts and insurance companies are investing in a number of oil and oil-related industries. In addition, banking interests from outwith Scotland, notably from London and North America, are steadily increasing their representation in Edinburgh, Glasgow and Aberdeen.

Apart from direct participation by Scottish financial interests in the offshore industry, there are several major fields in which Scottish-based companies outwith the oil industry have taken the initiative to a notable degree. These include substantial investments in the property market, the establishment or expansion of service companies and bases, the undertaking of major civil engineering contracts, participation in consortia for the construction of production platforms, and the establishment of module building yards. Finance for all such developments has also come from outwith the country, but the record of the Scottish institutions has certainly been good and was especially so in the earlier stages before the full potential of the market was established.

2. Labour Supply and Training

The impact of oil is often most immediately felt through its effect upon the labour situation. In the case of Scotland a good indication of the pattern of development may be had from Table 5.6 which shows the geographical distribution of employment created by oil-related industry. These figures have a number of outstanding features. First, a very large proportion (over 67%) of the total jobs in December 1974 was concentrated in two areas, namely Inverness and Easter Ross (the Cromarty and Moray Firths) and the North East (mainly the Aberdeen area and Peterhead). If the figure for the rest of the Highlands and Islands is included the proportion rises to over 77%. Secondly, the figures show that the relative distribution of jobs between regions has remained roughly the same over the past two years. Thirdly, despite a four-fold increase in the number employed in oil-related industries in two years, the total is less than 1% of the total insured population (over 2 million) in Scotland.

Although the figures from the Department of Employment are the best available it is likely they underestimate the true situation. It is relatively easy to obtain figures concerning the numbers employed by firms fully

TABLE 5.6 CHANGES IN OIL INDUSTRY EMPLOYMENT, MARCH 1973–DECEMBER 1974

	Mar. '73	Jun. '73	Sep. '73	Dec. '73	Mar. '74	Jul. '74	Oct. '74	Dec. '74
Inverness and Easter Ross	1 840	1 795	2 040	3 205	4 175	4 375	3 520	4 025
Remainder of Highlands and Islands	50	65	50	85	395	930	1 365	1 565
North East	1 410	2 305	2 305[b]	3 730	4 065	4 715	5 495	6 925
Tayside	25	35	95	135	150	280	475	765
East Central[a]	665	770	910	975	1 815	2 530	2 430	2 080
West Central[c]	110	170	250	480	675	785	855	870
Total	4 100	5 140	5 650	8 610	11 275	13 615	14 140	16 230

a. East Central is equivalent to the Firth of Forth region as described in Chapter 6.
b. No fresh figure was published for this quarter, therefore previous total has been carried forward.
c. The figures for West Central Scotland do not include workers engaged in rig construction work in Clyde shipyards, e.g. Marathon, Clydebank. In December 1974 workers in this category numbered 1935.

Source: Department of Employment.

engaged in oil-related work. There are a very large number of other companies, however, of which it is impossible to keep track, whose main activity lies elsewhere but which are engaged often to a significant degree in sub-contract work for specialist suppliers, especially in operational manufacturing and servicing facilities. Many of these firms are located in the North East and West Central regions, consequently the greatest shortfalls in the figures in Table 5.6 are likely to occur in those areas. Since the Department of Employment figures also exclude workers employed in rig and other construction work in Clyde shipyards such as Marathon (p. 104), this has the effect of further depressing the totals for both the West Central region and the country as a whole.

The greatest scope for oil-related employment has occurred in areas lacking a large reservoir of skilled industrial manpower. This has meant that the demand for labour has had to be met for the most part by attracting manpower from existing local industries or by drawing labour from other parts of the country and beyond. This greatly increased opportunity for employment and much higher incomes is welcome and has had the added advantage of widening the economic base both locally and regionally. However, the rapidity with which many of these changes have been brought about has often had serious implications for existing industry and has pinpointed deficiencies in the infrastructure of the areas most affected. The cyclical pattern of employment in projects such as platform fabrication and pipe-coating, associated with the immense size of individual contracts, has intensified these problems in some areas, particularly round the Cromarty Firth.

Competition for labour between oil-related industries and existing firms is often intensified by the higher basic wages paid by companies in the former category, coupled with opportunities for plenty of overtime. With overtime, wages of £100 per week for tradesmen are not uncommon. The effect is compounded by the fact that wages in the east, and especially the north, were often in the past below the rates paid by comparable industries elsewhere. The building industry has been particularly hampered by the loss of manpower to construction projects and fabrication work associated with the oil industry. Shipbuilding (p. 104) and boat building have been likewise affected while in the service sector losses of manpower may be more serious than they appear at first sight. For example, the loss of one or more mechanics may result in a garage ceasing to provide the services previously available to a local community.

Within Scotland, imported labour has gone some way towards alleviating this labour shortage. However, many incoming workers possess either specialised skills or belong to the mobile category associated with the offshore industry on a world scale. A large proportion of this last group is drawn from the USA and Continental countries, notably France, the Netherlands and Italy. Generally, however, there is a great need to create a much larger pool of skilled labour in Scotland (p. 104). Where this has been achieved locally, foreign labour has been steadily phased out (p. 78).

At this point it is worth considering the steps being taken by the Government and the oil industry to meet the training needs of workers employed both onshore and offshore. Following consideration of the recommendations concerning training and education made in the IMEG Report,[26] the Government set up a working party in 1973 to review the situation. Present Government policy on training and educational needs for the oil industry are based on the recommendations formulated by this working party. Some of the more important developments which have taken place since are as follows.

From the beginning of 1974 the whole field of training and employment services has been the responsibility of the Manpower Services Commission (MSC). The Training Services Agency (TSA), an executive arm of the MSC, is concerned with training (mainly technicians) to meet the needs of the oil industry. The Government is also interested in assisting training programmes undertaken both by individual companies and the Petroleum Industry Training Board (PITB). For example, the American drilling company, SEDCO, operates a training school for rig operators in Aberdeen while, also in Aberdeen, the PITB organise lectures and practical courses to give training in skills and safety procedures necessary offshore.

In addition, a wide range of courses at several levels and of varying lengths are on offer in further education centres including colleges of technology and universities. For example, both Robert Gordon's Institute of Technology in Aberdeen and Glasgow College of Technology, offer full-time one-year courses in offshore engineering while the former also provides courses in oil technology and oil-rig survival. Heriot-Watt (which has a centre for underwater technology), Glasgow and Strathclyde universities all offer various engineering courses with special relevance to the offshore industry, while Aberdeen and Strathclyde provide courses in petroleum geology and applied geology respectively. At time of writing, Dundee University is considering introducing a four-week course in hyperbaric diving techniques to meet offshore diving demands. The Government have also decided to set up a training centre for drilling at Livingstone (p. 101) and a school for deep sea diving at Loch Linnhe (p. 106).

Although Table 5.6 suggests that in terms of employment the West Central region has benefited little from oil-related industry, the real situation is probably a good deal healthier than mere statistics portray (see also (p. 102). Further, no figures are available as to the numbers of workers originating from West Central Scotland who are presently employed on oil-related work in other parts of the country and at sea. Nevertheless, there must be particular concern on economic and political grounds that of the 100 000 people currently (January 1975) unemployed in Scotland, a large proportion live within 40 km of the centre of Glasgow. It is, therefore, not surprising that the Government, particularly through the medium of the Offshore Supplies Office should be endeavouring to spread the direct employment effects of North Sea oil and gas more evenly throughout the country.[27]

The economic problems of the West Central region are, however, so deep-set, complex and on such a large scale, that oil-related employment is unlikely to make more than a token contribution to the region's unemployment difficulties. There are several basic reasons why this is so. First, as already mentioned, the total number of people employed in oil-related work is small and likely to remain so in relation to the size of the total insured labour force in Scotland.[28] Secondly, the largest direct employers of labour (platform yards) are by their nature excluded from the region although some yards on the Firth of Clyde, both operational and projected, can be serviced by labour from Clydeside. Thirdly, many types of oil-related industry have and will continue to have a market preference for an east coast location. To compel such industries to locate in the West Central Belt instead would weaken their competitiveness and encourage them to move elsewhere. Fourthly, the oil industry offers too narrow an industrial base to deal effectively with the industrial structural problems facing the region.

Undoubtedly, the main economic value of offshore oil and gas is not in the direct employment they provide, important though this may be, but rather in the revenues which flow through various forms of taxation once the fields are in production. In areas like the West Central region where the problem is basically an infrastructural one, the large scale reinvestment of revenues derived from oil and gas, if properly implemented, could do much to improve both environment and economic structure. At price levels pertaining in January 1975 it has been estimated that for West Central Scotland alone this may cost £10 000 million.[29]

3. Population and Services

Incoming populations tend to differ from local populations in terms of structure, notably age composition, the relative balance in numbers between the sexes and occupational groupings. In most oil-related industries employment opportunities for men are greater than for women, and an imbalance in the established pattern is likely to ensue as a preponderance of single males migrate into an area. The social consequences of this type of immigration are a subject of concern, particularly in small communities where the establishment of labour camps, not to mention the use of passenger liners anchored close inshore for accommodation (p. 80), are viewed as a threat to the social fabric. The difficulties are intensified by the fact that developments are often concentrated in small rural communities. It is precisely in such localities that problems of accommodation due to a serious shortage of housing are most acute.

The consequences of a sudden influx of population are apparent in the whole problem of housing provision and the availability of land. The latter is, of course, also pertinent to industrial development with the result that a very strong upward movement in land and property prices has generally taken place. In this situation industrial development has tended to take first priority, so that the housing shortage has become even more acute. For instance in the wider Aberdeen area, that is within a 30 km radius of the city centre, it has been estimated that 16 550 houses will be required between January 1972 and December 1976.[30] In 1973, the last full year for which building figures for this area are presently available, the building rate had slumped badly to only 742 houses per annum.[31] Since then both Aberdeen city and county have adopted non-traditional systems of building in order to speed up development, with effects which are apparent in Table 5.7. Of special note are the large numbers of houses under construction or approved at the end of 1974 in both the private and public sectors.

The organisation of house building can be divided into several categories. These comprise local authority housing financed by central government; private development, financed mainly by building society loans; and schemes built under the auspices of the Scottish Special Housing Association (SSHA). In the context of the oil industry, provision by the government and the SSHA is already assuming special significance, not least because such schemes are planned on a large scale.

For example, the SSHA plans to build 4 000 houses in the area between Dundee and Caithness over the period 1973–78.[32] Nearly one-half of this programme was under construction at the beginning of 1975, with some houses already completed on eleven sites. The need for homes has been so great in some places, that in addition to the houses designed and built by them, the Association has bought groups of houses from private builders. To date the housing shortage has led to situations in which local people wishing to purchase homes have often found themselves priced out of the market. In 1974 the high rate of inflation and difficulties incurred in obtaining mortgages led to some slackening in demand. Although houses were taking longer to sell, however, this did not result in any significant fall in prices, especially in those areas where the general shortage was greatest. Even though the housing outlook seems more favourable in 1975 than it has been for some time the shortage of houses remains a restricting factor in the restructuring of the economy of Scotland to cater for the oil industry.

The provision of housing over the past two years in the three areas most affected by oil-related development is given in Table 5.7.

Apart from the problem of housing development, population growth brings strong pressures to bear on public and private services. In the public sector new requirements must be met, especially in the fields of water supply and drainage, roads, electricity, health and welfare, and education. For example, schemes to boost water supplies are already planned or in progress, while new schools will be needed. However, there has been a substantial time-lag in improving roads to meet the needs of the oil industry, although some much needed reconstruction is now in progress (p. 52).

In the private sector special pressures have been placed on the hotel industry. On the one hand a big expansion in hotel accommodation has already taken place and more is planned, especially in the Aberdeen area. On the other, the demands for accommodation created by oil are conflicting with the needs of the tourist industry, notably in Shetland and Easter Ross, where hotel accommodation is often limited but where the tourist industry has traditionally been one of the mainstays of the economy.

The increase of population, the expansion of existing industries and the introduction of new industries put pressures on the quality of life in ways which are not easily measured. The speed and sense of urgency of developments are two of the most subtle influences which the new discoveries are exerting in those parts of Scotland where a rather leisurely pace of life has been one of the most attractive characteristics. Whether this quality of life will be maintained in such areas is open to debate and is a frequent topic of concern at public inquiries into applications for planning permission for new developments, such as are discussed in Chapter 7.

TABLE 5.7   PROVISION OF HOUSING IN THE THREE MAIN OIL-AFFECTED AREAS, 1973 AND 1974

	House Completions				Houses under Construction at the end of 1974	
	Public Sector		Private Sector		Public Sector[a]	Private Sector
	1973	1974	1973	1974		
Shetland	90	71	64	109	300	200
Cromarty and Moray Firths	300	1 079	341	295	2 300	700
Aberdeen City and County	861	917	564	1 115	4 500	1 400
Total	1 251	2 067	969	1 519	7 100	2 300

a Includes houses approved and awaiting a start.

Source: North Sea Oil Division, Scottish Economic Planning Department

REFERENCES AND NOTES

1. Compare for example the figures quoted in the International Management and Engineering Group (IMEG) (1973), *Study of potential benefits to British Industry from offshore oil and gas developments*, HMSO, London, and the Department of Trade and Industry (1973), *North Sea Oil and Gas: a report to Parliament*, HMSO, London. (Reprinted from *Trade and Industry*. **10**, No. 4, 25 January, 1973).

2. ODELL, P. R. (1974), 'The Oil Dimension', *Investing in Scotland's Future*, Fifth International Forum, Aviemore, 6–7 November, 1974.

3. BEXON, R., (1974), *The North Sea—Achievements and Prospects*, International Offshore Technology Conference, London, 9–10 October, 1974.

4. ODELL, P. R. (1974), *op. cit.*

5. Department of Energy (1974), *Production and reserves of oil and gas in the United Kingdom: a report to Parliament*, HMSO, London, May 1974.

6. *North Sea Oil Information Sheet: February 1975*, NSOIS 75 (1). North Sea Oil Division, Scottish Economic Planning Department, Edinburgh, 20 February 1975.

7. Date and estimated production rate given by ELF (Elf Oil Exploration and Production Ltd) and later updated—see CUMMINGS, B. (1975), 'Frigg gas hold-up,' *The Press and Journal*, 23 January, 1975. Production rate is approximately equivalent to 2 000 million ft^3/day.

8. BAUR, C. (1971), 'Country's oil consumption likely to reach 20m tons by 1980,' *The Scotsman*, 2 July, 1971.

9. BAUR, C. (1972), 'Forties Field will lead to massive rise in oil exports from Forth,' *The Scotsman*, 22 March, 1972.

10. FRAZER, F. (1974), 'Oil bonus could increase flow to Forth,' *The Scotsman*, 12 November, 1974.

11. Project Independence, introduced by the Nixon administration in response to the Arab oil blockade of 1973–74, is designed to free the US from foreign energy sources by 1980. HAWKES, N. (1974), 'Strip-mining threatens America the Beautiful,' *The Sunday Times*, 31 March, 1974.

12. MACKAY, D. & MACKAY, A. (1975), 'Scotland on Stream—whose hand is on the oil wealth?' *The Scotsman*, 7 February, 1975.

13. HOLS, A. (1974), 'How to evaluate the economics of North Sea oilfields—Part I,' *Petroleum International*, London, November, 1974, p.50.

14. BP (1975) *Forties News*, February, 1975.

15. BP (1973) *Forties News*, May, 1973.

16. BP (1975), *op. cit.*

17. BP (1975), *op. cit.*

18. Based on a table in BUCKMAN, D. & CRANFIELD, J. (1974), 'North Sea card filling up fast,' *Petroleum International*, London, November, 1974, p.23.

19. 'Oilmen invest massive stake,' *The Press and Journal*, 12 February, 1975, quoting the latest edition of the Chase Manhattan Bank's authoritative *Capital investments of the world petroleum industry*.

20. Information supplied by BP Petroleum Development Ltd.

21. The arcs on Fig. 5.2 have been drawn so as to allow for the way in which passage round headlands and islands reduces the range of ships in certain directions.

22. Individual towns involved are named in capitals in Figs. 6.5, 6.6, 6.11 and 6.16.

23. A platform being built currently at Rotterdam by ANDOC is also expected to be towed to the Inner Sound for completion.

24. The Salveson group of Leith has, however, a drill-ship in operation off Australia and has also entered the specialist field of well-casing engineering.

25. For instance, in 1973 the Bank of Scotland together with the Republic National Bank of Dallas took the initiative in founding the International Energy Bank, a consortium of British, US, Canadian and French banks which finances particularly large energy projects. All the Scottish clearing banks now have offices in the USA to maintain on-the-spot'links with the oil industry and associated financial interests.

26. IMEG Report (1973), *op. cit.* (1 above)

27. See also Scottish Development Department (1974), *North Sea Oil and Gas Coastal Planning Guidelines*, Scottish Development Department, Edinburgh, August 1974. This document sets out various guidelines for local authorities and prospective developers regarding the location of oil-related activities in Scotland, and distinguishes between preferred development zones and preferred conservation zones.

28. MACKAY, D. I. (1975), *North Sea Oil and the Scottish Economy, North Sea Study Occasional Papers No. 1*, University of Aberdeen, Department of Political Economy, January 1975 (p. 10). Professor Mackay estimates that direct employment in operational activities and related services, the production of equipment and materials, the processing of oil and gas and associated industrial developments may amount to 25 000–30 000 when such employment reaches its peak in the late 1970s.

29. From a speech by Gordon Wilson, M.P., in the debate on devolution in Parliament, 3–4 February, 1975, *Hansard, Vol. 885, Nos 59 and 60*, HMSO, London.

30. DUNBAR, J. (1975), 'Aberdeen housing shortfall—but North East is faring better,' *The Press and Journal, Construction Review '75*, 18 February, 1975.

31. DUNBAR, J. (1975), *op. cit.* (30 above).

32. GOURLAY, M. (1975), 'A 4 000-house programme,' *The Press and Journal, Construction Review '75*, 18 February, 1975.

6 Regional Studies

Elements of the regional distribution of the oil industry have already emerged in Chapter 5, but understanding of how varied are the nature and scale of developments in different parts of the country, and of how widely different is their impact on local communities, becomes apparent only when they are examined at regional and local levels.

Most of the impact of the offshore industry is in fact highly localised, and it has been felt much more in lightly populated and less industrialised areas than in the crowded Central Lowlands. Changes in land use, population and economic structure have been longest established and have gone furthest in the Aberdeen district and around the Cromarty Firth. Developments came rather later to Peterhead and Shetland, but their pressure has been equally precipitant and intensive. In Orkney and Wester Ross isolated but dramatic changes are in hand at individual sites, and similar schemes are starting or are in prospect in Lewis and on the Firth of Clyde. By contrast, however, Tayside and the Firth of Forth have had considerably more time to accommodate developments, and are somewhat disappointed that demand for space and facilities has not been more brisk. In West Central Scotland the public at large has so far scarcely noticed the impact of the oil industry but the heavy industries of the area are increasingly involved in meeting its engineering needs.

The proportions of space in this chapter devoted to individual regions reflect their relative importance, but the sequence in which these are treated is determined by their location rather than their function. In so far as a functional distribution pattern has emerged to date, the distinction is between those areas primarily concerned with servicing offshore operations and providing landfalls for pipelines, that is in Shetland and Orkney, the North East and Tayside; and those involved in or projected for platform construction and pipeline preparation, namely the Forth, Moray and Cromarty firths, Wester Ross, Lewis and the Clyde. Additionally, Aberdeen has become the hub of the industry as a whole, and the Forth still the only clearly determined centre for processing the oil obtained.

While physical factors of site and location relative to concession areas have obviously been important in establishing this pattern of growth (Fig. 5.2), the readiness of communities to respond to the challenges offered has also been a considerable influence. Where there have been local disagreements about the desirability of developments, delays and, in some cases, changes in the eventual location pattern of the industry have resulted. Chance elements have also entered the location pattern especially where, as so often, needs have been met by improvising or altering the uses of existing facilities. Examples include the virtually unused Peterhead Harbour of Refuge, which has become a very active port, and the former Wellesley Colliery at Methil, now a platform building yard.

The maps in this chapter show the location of the oil industry at the end of January 1975. These have been based on both published and unpublished material obtained from the Press, local authorities, harbour boards, oil companies and companies serving the oil industry, but some, such as Figs. 6.2, 6.7 and 6.8, have in large part depended on fieldwork by the authors. It should be noted that despite the degree of detail in Fig. 6.8 it omits a considerable number of local firms and established branches of national companies which draw custom from the oil industry. This was done both because it proved too difficult to assess how substantially they are committed to the industry, and also to concentrate attention on the pattern of changes in land use which have occurred because of the oil 'boom'.

1. Shetland

The importance of physical accessibility is nowhere more clearly illustrated than in Shetland, which occupies a strategic location relative to existing and potential discoveries in the northern part of the continental shelf, both in Scottish and Norwegian waters (Fig. 5.2). It also has natural harbours suitable for the establishment of bases and terminals. With the exception of Lyness in

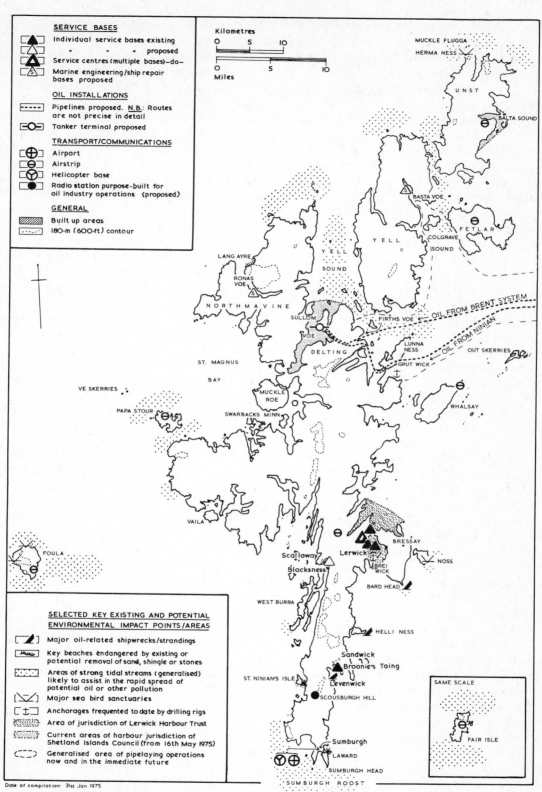

SERVICE BASES

Individual service bases existing
" " proposed
Service centres (multiple bases) –do–
Marine engineering/ship repair
bases proposed

OIL INSTALLATIONS

Pipelines proposed. N.B.: Routes
are not precise in detail
Tanker terminal proposed

TRANSPORT/COMMUNICATIONS

Airport
Airstrip
Helicopter base
Radio station purpose-built for
oil industry operations (proposed)

GENERAL

Built up areas
180-m (600-ft) contour

Kilometres
0   5   10

Miles
0   5   10

**SELECTED KEY EXISTING AND POTENTIAL
ENVIRONMENTAL IMPACT POINTS/AREAS**

Major oil-related shipwrecks/strandings
Key beaches endangered by existing or
potential removal of sand, shingle or stones
Areas of strong tidal streams (generalised)
likely to assist in the rapid spread of
potential oil or other pollution
Major sea bird sanctuaries
Anchorages frequented to date by drilling rigs
Area of jurisdiction of Lerwick Harbour Trust
Current areas of harbour jurisdiction of
Shetland Islands Council (from 16th May 1975)
Generalised area of pipelaying operations
now and in the immediate future

Date of compilation: 31st Jan 1975

FIG 6.1   SHETLAND

SAME SCALE

FAIR ISLE

MUCKLE FLUGGA
HERMA NESS
UNST
BALTA SOUND
BASTA VOE
FETLAR
COLGRAVE SOUND
YELL
YELL SOUND
OIL FROM BRENT SYSTEM
LANG AYRE
RONAS VOE
NORTHMAVINE
SULLOM VOE
FIRTHS VOE
OIL FROM NINIAN
LUNNA NESS
OUT SKERRIES
DELTING
GRUT WICK
ST. MAGNUS BAY
MUCKLE ROE
WHALSAY
VE SKERRIES
PAPA STOUR
SWARBACKS MINN
VAILA
BRESSAY
NOSS
Scalloway
Lerwick
Blacksness
BREI WICK
BARD HEAD
FOULA
WEST BURRA
HELLI NESS
Sandwick
Broonie's Taing
ST. NINIAN'S ISLE
Levenwick
Scousburgh Hill
Sumburgh
LAWARD
SUMBURGH HEAD
SUMBURGH ROOST

Orkney (p. 73), all the 'forward' bases for rig-servicing are located in Shetland and the largest oil terminal projected to date is planned for Sullom Voe. Recognising from an early stage the dangers of inadequately controlled oil developments on their island community, Zetland County Council pioneered a distinctive approach to planning legislation. Its concern to obtain greater control of events than would otherwise be possible under normal planning legislation foreshadowed, at least in part, the actions of two successive UK governments.

There are four main aspects of development in Shetland to be considered: the oil industry itself; the expansion of service activities and their relationship to local business enterprise; the impact on the environment, economy and community; and the role of the local authority, especially in relation to the development of the Sullom Voe terminal.

An important result of the extremely successful exploration to date of the East Shetland Basin has been the conception of an integrated production system for many of the large fields discovered (Fig. 4.6). The Brent System is now in its first stages of implementation, with some platform orders having been placed and preliminary work having commenced close to the east coast of the islands for bringing the first pipeline ashore to Sullom Voe. Also, a second production system linking the Ninian and Heather Fields by a separate pipeline to Sullom Voe is in the advanced planning stages, with platform orders being placed and survey work on the pipeline route in progress. Meanwhile, there remains a strong possibility that oil will be piped to Shetland from the Norwegian Sector, notably from the recently discovered Statfjord Field. Latest estimates indicate that, by 1980, approximately 60% of the North Sea production from proved fields in the East Shetland Basin will be landed in Shetland (Table 5.3). It is against such a background that the enormous developments which are about to inundate the islands—especially the Sullom Voe Terminal complex—must be viewed.

Meanwhile, exploratory activity continues to build up. Of the 20 or so rigs operating in the vicinity of the islands in the summer of 1974, nearly all were drilling in the East Shetland Basin (Table 4.3). Although Aberdeen remains the main base for ships servicing these rigs, a number of them are also being serviced from Shetland and this activity is evidenced on land by the continuing expansion of the service industry. The locations of existing and proposed service bases are indicated in Fig. 6.1. These correspond in most cases to the major centres of the great herring fishing period from the 1880s until the 1930s, which had similar requirements

for extensive shore facilities, particularly flat land and large harbours.

Prior to 1974, servicing activity was largely carried out at a relatively small scale from pre-existing premises in the Lerwick Harbour area, together with a very little from Scalloway. With the construction of purpose-built bases at the Green Head and Holmsgarth, at Lerwick (Fig. 6.2) and at Broonie's Taing at Sandwick (Fig. 6.1), this service capacity is rapidly building up, even although construction is not yet complete in some cases. For example, in the course of 1974 some 838 oil-related shipping movements were recorded in Lerwick Harbour, compared with 328 in 1973.[1] Planning permission for a total of 31 supply ship berths has been granted in Shetland, and further planning applications have been lodged for bases in the Lerwick Harbour area. It is worth noting that the bases proposed for Basta Voe and Ronas Voe (Fig. 6.1) are intended to be ship repair and marine engineering centres rather than conventional supply bases. Basta Voe is likely also to become a pipe storage centre. Meanwhile, there has been a re-organisation of local marine engineering firms, with new companies being formed consisting of both Shetland and Scottish mainland engineering interests. These continue to cater primarily for the fishing industry, but have an eye to expansion associated with the build-up of oil-related work generally, including acquisition by local firms of agencies for suppliers of equipment outwith Shetland.

In the first edition of *Scotland and Oil*,[2] it was noted that the visual impact of the oil industry on Shetland in the summer of 1973 was minimal. At sea, apart from the comings and goings of a few supply vessels, interest focused on the misfortunes of oil-related shipping. For example, a barge carrying 700 tons of installation jackets for the Ekofisk Field had been lost at the foot of Bard Head, within two miles of the nationally famous bird sanctuary of Noss (Fig. 6.1). Since then there has been a further serious wreck, namely, the stand-by vessel *Greyfish* on Helli Ness (Fig. 6.1). Additionally, there has been a continuing sequence of minor incidents in the form of strandings of vessels due to bad weather or navigational errors, in many cases potentially serious.

By the end of 1974 the impact of oil on land, environmental and otherwise, had become marked. Large scale reclamation associated with the building of service bases was continuing, especially in Lerwick Harbour where major schemes, including the 12 ha (30 ac) Norscot base, were at an advanced stage (Fig. 6.2). The construction of a camp for the Sullom Voe terminal complex had begun and test tunnels were being driven into Calback Ness to ascertain its suitability for the underground storage of

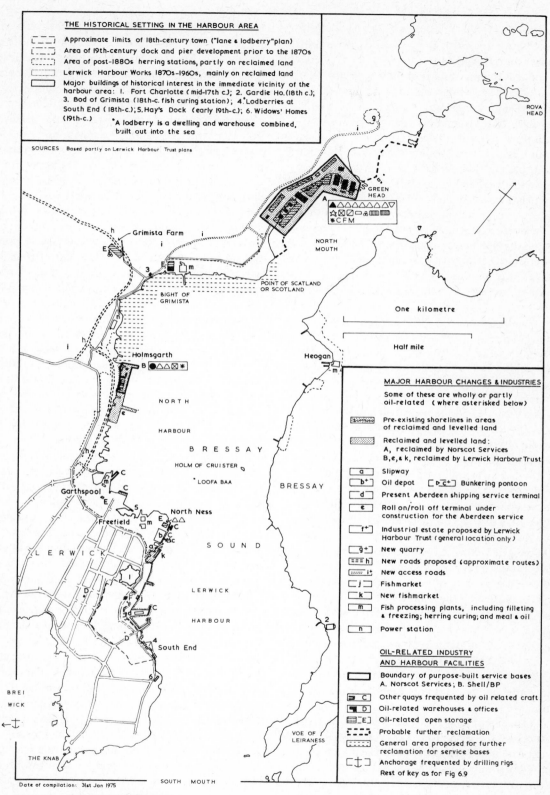

THE HISTORICAL SETTING IN THE HARBOUR AREA

Approximate limits of 18th-century town ("lane & lodberry" plan)
Area of 19th-century dock and pier development prior to the 1870s
Area of post-1880s herring stations, partly on reclaimed land
Lerwick Harbour Works 1870s–1960s, mainly on reclaimed land
Major buildings of historical interest in the immediate vicinity of the harbour area: 1. Fort Charlotte (mid-17th c.); 2. Gardie Ho. (18th c.); 3. Bod of Grimista (18th-c. fish curing station); 4. Lodberries at South End (18th-c.); 5. Hay's Dock (early 19th-c.); 6. Widows' Homes (19th-c.)

*A lodberry is a dwelling and warehouse combined, built out into the sea

SOURCES Based partly on Lerwick Harbour Trust plans

Grimista Farm
ROVA HEAD
GREEN HEAD
NORTH MOUTH
Point of Scatland or Scotland
BIGHT OF GRIMISTA
Holmsgarth
Heogan
NORTH HARBOUR
B R E S S A Y
HOLM OF CRUISTER
LOOFA BAA
B R E S S A Y
One kilometre
Half mile
Garthspool
North Ness
Freefield
L E R W I C K
S O U N D
LERWICK HARBOUR
South End
BREI WICK
THE KNAB
VOE OF LEIRANESS
SOUTH MOUTH

Date of compilation: 31st Jan 1975

MAJOR HARBOUR CHANGES & INDUSTRIES
Some of these are wholly or partly oil-related (where asterisked below)

Pre-existing shorelines in areas of reclaimed and levelled land
Reclaimed and levelled land: A, reclaimed by Norscot Services B, e, & k, reclaimed by Lerwick Harbour Trust
a   Slipway
b+ Oil depot   ▷c+ Bunkering pontoon
d   Present Aberdeen shipping service terminal
e   Roll on/roll off terminal under construction for the Aberdeen service
f+ Industrial estate proposed by Lerwick Harbour Trust (general location only)
g+ New quarry
h   New roads proposed (approximate routes)
i+ New access roads
j   Fishmarket
k   New fishmarket
m   Fish processing plants, including filleting & freezing; herring curing; and meal & oil
n   Power station

OIL-RELATED INDUSTRY AND HARBOUR FACILITIES

Boundary of purpose-built service bases A. Norscot Services; B. Shell/BP
C   Other quays frequented by oil related craft
D   Oil-related warehouses & offices
E   Oil-related open storage
Probable further reclamation
General area proposed for further reclamation for service bases
Anchorage frequented by drilling rigs
Rest of key as for Fig 6.9

FIG 6.2   LERWICK HARBOUR

69

crude oil (Fig. 6.3). Preparatory work for the laying of the Brent pipeline was also going ahead in the eastern approaches to Yell Sound.

The economic impact of oil is now also becoming evident. All kinds of labour, especially skilled labour, is in very short supply. High wages paid by oil-related industry have attracted many workers from other occupations, at considerable disadvantage in particular to the public services, where a system similar to 'London weighting' has been sought for wages. Of the basic industries, the woollen industry has been most vulnerable to labour shortages. Agriculture has perhaps been little affected in terms of loss of manpower, and few fishermen have gone over to oil-related employment. Most agricultural holdings in Shetland are run as family units, while until comparatively recently the fishing industry has been very prosperous. Needless to say, accommodation of all kinds is desperately short and house prices extremely high.

Communities, and groups within them, have responded in differing ways to the changes arising from the advent of the oil industry. Since much development remains pending rather than in existence, issues have tended to centre round four major themes; industrial development, use of land and sea, conservation, and the consequences for island life. Sensitivity to the vulnerability of the traditional basic industries, especially fishing and knitwear—the latter particularly subject to fluctuations in demand—has already encouraged substantial and perhaps increasing support for industrial development among the public at large.[3] This has been most clearly illustrated to date by the referendum organised early in 1973 by the Scalloway Development Committee in which a narrow majority of the village population were in favour of the establishment of a large service base which would have radically altered the character of the district. More recently, a strong desire has been expressed by the majority of Yell people, notably through Yell Community Association, for the establishment of the Basta Voe base, on the grounds that knitwear and fishing are in decline, reinforced by the arguments that Yell has always been one of the more economically depressed areas in Shetland and consequently among the first to feel the effects of a recession.

The use of land and sea has been brought into sharp focus by the scale of physical planning required to accommodate developments of such magnitude as the Sullom Voe terminal, and is in practice closely bound up with conservation issues (Figs. 6.1, 6.3). These problems are perhaps least acute in Lerwick, where the large

harbour area is well-suited to industrial expansion without further physical encroachment upon the existing community. The harbour already bears witness to former extensive industrial development in the shape of many areas of derelict herring stations dating from the heyday of the herring fishing prior to 1939. These are currently being obliterated by reclamation schemes associated with the building of service bases and a roll-on/roll-off ferry terminal (Fig. 6.2). Elsewhere, planning problems are often more acute. In Scalloway, for example, there is insufficient room for the scale of development taking place at Lerwick, while the Broonie's Taing base is built on a relatively exposed coast. In rural crofting areas, notably around Sullom Voe and in Unst, the immediate issue has been the compulsory purchase of land, which led in the first instance to the majority of local people being opposed to County Council policy directed towards controlling development, including the speculative activities of property companies. With increasing realisation of the implications of development, however, this attitude has been largely replaced by one of support for Council policy.

Meanwhile, at sea, apart from the problems of acquisition of harbour authority (see below), clashes between fishing and oil interests have taken place, albeit on a relatively minor scale. Already there have been incidents associated with oil-related work in the Yell Sound area involving the use of piers and the presence of inadequately lighted buoys. Of more serious long-term import, however, has been disturbance of inshore shell fisheries by oil-related craft engaged in preparatory work for the Brent pipeline. Such conflicts are inevitable, as projected pipeline routes crossing inshore waters are liable to result in partial destruction of the major shell fish beds in Yell Sound and immediately to the east. These circumstances have already resulted in refusal by the Council of planning permission for a temporary floating base to service the Brent pipeline project, on the grounds that pollution from the base might harm shell fish grounds. In any case pipelines, especially if they cannot be buried (p. 23), will tend to 'sterilise' fishing grounds due to the dangers of fishing too close and losing gear through fouling the pipelines. It is for this reason mainly that strong opposition was mounted by the Shetland fishermen to the proposed crossing by the Brent pipeline of the important Pobie Bank, a leading white fishing area east of Unst. It should be pointed out, however, that there has been consultation throughout between the oil companies and the Shetland Fishermen's Association aimed at sorting out such problems as they arise.

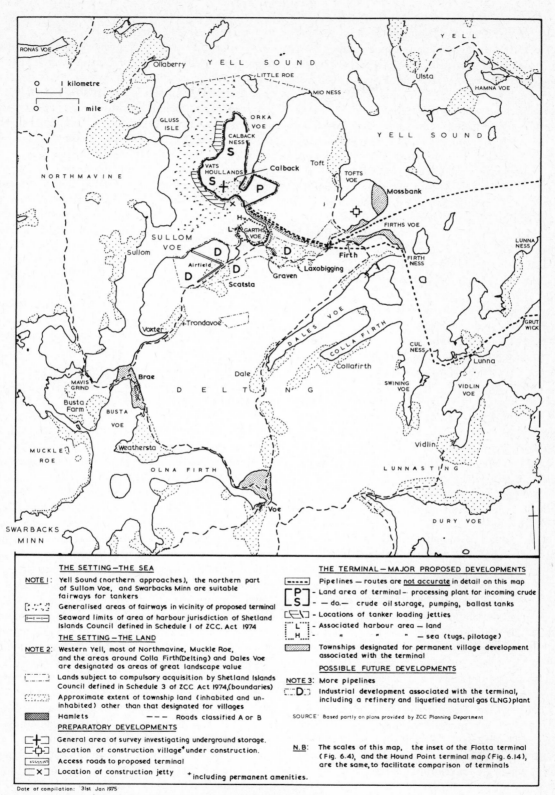

THE SETTING—THE SEA

NOTE I: Yell Sound (northern approaches), the northern part of Sullom Voe, and Swarbacks Minn are suitable fairways for tankers

Generalised areas of fairways in vicinity of proposed terminal

Seaward limits of area of harbour jurisdiction of Shetland Islands Council defined in Schedule I of ZCC. Act 1974

THE SETTING—THE LAND

NOTE 2: Western Yell, most of Northmavine, Muckle Roe, and the areas around Colla Firth (Delting) and Dales Voe are designated as areas of great landscape value

Lands subject to compulsory acquisition by Shetland Islands Council defined in Schedule 3 of ZCC Act 1974, (boundaries)

Approximate extent of township land (inhabited and uninhabited) other than that designated for villages

Hamlets   — — — Roads classified A or B

PREPARATORY DEVELOPMENTS

General area of survey investigating underground storage.

Location of construction village* under construction.

Access roads to proposed terminal

Location of construction jetty   *including permanent amenities.

THE TERMINAL—MAJOR PROPOSED DEVELOPMENTS

Pipelines — routes are not accurate in detail on this map

[P] Land area of terminal — processing plant for incoming crude
[S] —do.— crude oil storage, pumping, ballast tanks

Locations of tanker loading jetties

L - Associated harbour area — land
H -   "   "   "   - sea (tugs, pilotage)

Townships designated for permanent village development associated with the terminal

POSSIBLE FUTURE DEVELOPMENTS

NOTE 3: More pipelines

D Industrial development associated with the terminal, including a refinery and liquefied natural gas (LNG) plant

SOURCE: Based partly on plans provided by ZCC Planning Department

N.B: The scales of this map, the inset of the Flotta terminal (Fig. 6.4), and the Hound Point terminal map (Fig. 6.14), are the same, to facilitate comparison of terminals

Date of compilation: 31st Jan 1975

FIG 6.3   SULLOM VOE AND ADJACENT DEVELOPMENTS

A further danger to fishing, as in other spheres, is that of pollution from pipeline fractures, tanker operations and other causes. This was highlighted by the major spillage at the Bantry Bay Terminal in 1974, the results of which were witnessed by some Shetland Council representatives on a fact-finding visit. Considerable environmental problems are also posed on land by the needs for sand and rock of a vastly expanded construction industry. This gives rise to acute problems of beach conservation, notably at the St. Ninian's Isle tombola, the Lang Ayre cliff-foot beach on the seaward side of Ronas Hill, and the Laward storm beach near Sumburgh Airport (Fig. 6.1). Opening up of new quarries can result in a serious loss of amenity for people living in the immediate vicinity, and domestic water supplies require safeguarding against over-use by industrial concerns. This is exemplified by the problems being encountered in securing quarries and water supplies for the runway improvements being undertaken at Sumburgh Airport. Shetland Civic Society is particularly concerned that the best of the old should be conserved, and is also actively opposing any future plans for building a refinery or liquefied natural gas plant in association with the new terminal at Sullom Voe. Meanwhile, a special environmental advisory group including members of the County Council, oil companies, Aberdeen and Dundee universities, the Nature Conservancy and Countryside Commission, has been set up to monitor the developments at Sullom Voe; and a Shetland branch of Friends of the Earth is also getting under way.

Whether oil is considered to be a good thing or a bad thing for Shetland ultimately depends on the set of values used in judgment, both of island life, and the effects upon it of oil developments. For those who consider that, on balance, oil is a bad thing, there is plenty of evidence to support their arguments. This includes a rising crime rate and a rapid build-up of an itinerant work force followed by permanent settlement by a large 'alien' population with a different set of social values, likely to bring about rapid social change and disruption of the island way of life. Shetland's population has already increased by about 1 000, from just over 17 500 at the 1971 Census, to around 18 500 at the end of 1974. There are also acute labour shortages threatening to undermine seriously basic industries and public services; house prices of the same order as those obtaining in Aberdeen and London; and roads (especially around Lerwick, and from Lerwick to Sullom Voe and Scalloway) which are deteriorating rapidly under the impact of heavy traffic unforeseen when they were made years ago; as well as the imminent despoilation of the

environment already discussed.

On the other hand, many people emphasise the potential benefits of oil developments, including greater material prosperity for the islands as a whole; increased opportunities for employment, notably for persons with professional and technical qualifications; opportunities to reverse population decline by encouraging expatriates to return; greater variety of opportunities in employment for unskilled and semi-skilled labour, especially in construction work; improved air and sea services; and improved telecommunications links, such as the new micro-wave radio links soon to be established. However, it is often pointed out that, unlike the traditional industries of agriculture, fishing and knitwear which, variously organised, have been the mainstays of the island economy for hundreds of years, oil will at best last only for a generation or two, and conceivably leave the islands with the same problems that have existed since the beginning of the century and from which they have been actively trying to escape since the Second World War. Among public bodies, Shetland Council of Social Service has been particularly active in warning the community of the social dangers inherent in the coming of oil.

The reaction of the local authority to all this has been to try and obtain a measure of long-term control of events for the benefit of Shetland as a whole. From having been behind the average local authority in planning matters—Zetland County Council having been one of the few local authorities in Scotland not to have prepared a County Development Plan under the 1947 Town and County Planning Act and, until recently, having neither a County Planning Officer nor a County Planning Department—the Council have pioneered ways in which small local authorities in remote rural areas can cope with massive industrial development. In order to accomplish this it has proved necessary to move beyond the powers available under the usual planning framework, and to acquire powers through an Act of Parliament which allows the Council actively to participate in developments.[4] One of the first steps was to produce an interim county development plan pending the preparation of a full plan, and to commission a planning study of the Sullom Voe—Swarbacks Minn area in preparation for the projected terminal facilities.[5] Preparing for this scheme has become one of the central preoccupations of the local authority, due to its enormous scale in comparison with other aspects of oil-related development.

The Zetland County Council Act 1974, put at its simplest (pp. 113—114) gives the Council powers to control the development of land and sea areas; to develop and operate harbour facilities, notably the

Sullom Voe terminal; and to apply financial benefits gained to the general good of the community as well as for the maintenance of the harbour undertaking. In order to accomplish this, the Zetland Finance Company Ltd has been created by the Council to borrow money on a large scale to build the terminal, and the Council has to date (January 1975) entered into partnership with commercial concerns in new companies which will provide tuggage and build a construction village for the terminal.

Close co-operation between the Council and the oil industry has been maintained throughout all these preliminary organisational phases of development. An agreement has been drawn up between the industry and the Council to compensate the islands for disturbance caused by the incoming industry, and half a million pounds has been handed over to the Council by the oil companies acting in concert, as a first instalment of future payments. Further, an Oil Liaison Committee consisting of the Council and companies operating in the offshore area has been established for some time as a consultative group. A Joint Planning Group consisting of each operator of pipelines to Shetland together with members of the Oil Liaison Committee has been set up for planning the terminal. Their plan has now been accepted, with modifications, by the Council, and the construction of the biggest oil terminal of its kind in the UK is due to commence in 1975.

2. Orkney

Developments in Orkney have paralleled in many ways those in Shetland, but they are on a much smaller scale reflecting both the more limited exploration work done so far off Orkney and the smaller number of discoveries made. Although well-placed in the long run to act as a servicing centre, especially for waters to the west of the islands (Fig. 5.2), Orkney's servicing function to date has been limited to a small advance base, opened in the autumn of 1973, on 1.6 ha (4 ac) of the former naval base at Lyness on Hoy. A consortium of BP, Shell and SEDCO has plans for development there, but so far the base has been used only intermittently by Shell Expro when there have been rigs drilling to the west of Orkney. Aberdeen-based helicopters operate out of Kirkwall airport as required, to maintain communications with the rigs, and Sea Mud Services provide service facilities.

Plans for a much larger service base, with an associated steel fabrication plant, at Carness near Kirkwall were put forward in 1972 by Hudson's Offshore Ltd. Despite buying an option on the land,[6] the company has not, however, applied for permission to develop the site. It would appear that the sequence of developments to the east of Shetland has made the Shetland Islands a much more attractive site to Hudson's, as to other service companies, over the past three years.

An early interest in Scapa Flow as a centre for platform construction has also faded. A UK-Danish company, Christiani and Nielson, put forward plans in October 1972 for building concrete platforms at the Bay of Houton on Mainland east of Stromness, but withdrew these proposals in March 1974. There had been considerable opposition locally, particularly to the associated plan to quarry large quantities of stone from an adjacent hill, but the reasons given by the company for abandoning the scheme were that they now considered the depths of water in Scapa Flow to be insufficient for the later stages of platform building.

Paradoxically, in view of the limited exploratory drilling which has been done in blocks accessible from Orkney, the only large-scale development which has been proceeded with in the islands is one for a pipeline landfall, oil treatment and shipment facilities for oil from the Piper Field. Orkney's attraction for this scheme is the sheltered water of Scapa Flow, in which large tankers will have both adequate depth and space to manoeuvre as they moor at the loading towers (Fig. 6.4). Developments in Scapa Flow and on the Golta Peninsula of Flotta have been in hand since March 1974. The first stage of the scheme should be complete by the summer of 1975, although oil will not come ashore until the following year (Table 5.3).

The pipeline for incoming crude oil has already been laid (January 1975) from Flotta to South Ronaldsay, and from just east of that island to Piper. Construction has started of two single point tanker mooring towers located about 2.8 km ($1\frac{3}{4}$ ml) north of Flotta. Two pipelines have been laid from the island to each of these, a 48″ pipe for the loading of oil and a 36″ pipe for taking of ballast water from the ships. Initially about ten 50 000 ton tankers a month are expected to carry oil from this terminal to Canvey Island on the Thames Estuary, where Occidental, the company developing the Piper Field, are building a refinery. At a later stage the Flotta terminal could be handling 200 000 ton vessels.

On Flotta itself the 93 ha (230 ac) site includes five oil storage tanks, holding a total of $2\frac{1}{2}$ million barrels, which are being constructed by the Motherwell Bridge Company. Plant to remove gas from the oil to

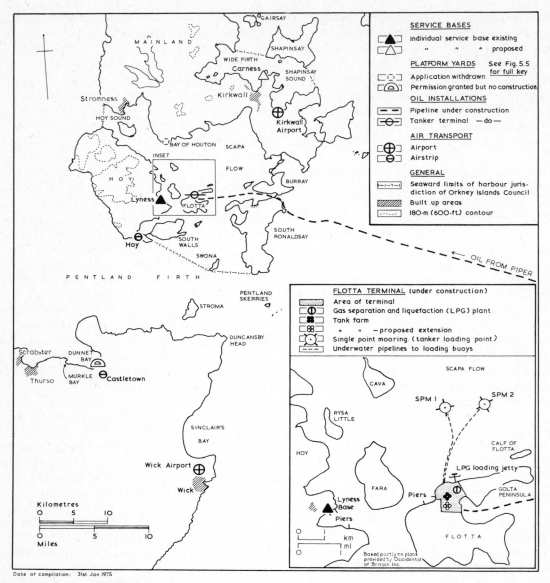

FIG 6.4   ORKNEY AND CAITHNESS

stabilise it for shipment by tanker is being built alongside by another company, also a gas liquefaction unit with a capacity of 4 000 barrels a day. Storage tanks for liquid gas and a loading jetty from which it will be shipped by coastal tankers complete the project presently in hand, but an application for the construction of three further oil tanks has just been announced (January 1975).

This application is a reminder that at all stages in the pipeline and terminal development the Occidental company, as agents for the companies owning the Piper

Field, has made provision for handling double the output initially projected to come from that field. Those units not already large enough can be readily expanded. Discovery of the Claymore Field by the same group in the summer of 1974, and generally increased interest recently in prospecting blocks adjacent to Piper, have justified this provision. Substantial expansion of the throughput at Flotta could, however, put further stress on a small community which is already under considerable strain.

In the short term, a construction force of up

to 500 men has had to be accommodated in a camp site specially constructed for them. In the longer term, between 80 and 100 permanent jobs are expected to be created for the running of the plant—all on an island which before development had a resident population of only around 80. The company policy is to employ Orkney men wherever possible, and the job opportunites are welcome in a community which previously was being steadily drained of its younger and more active people. But the greatly increased population and movement of people induced by the creation of new employment mean a need for rapid provision of housing, construction of an adequate airstrip and improvement of the piers and shipping serving the island. Progress in these matters has been much less spectacular so far than that made in developing the oil terminal. The local community has been feeling the strain and has protested to the County Council.[7]

The people of Orkney, like those of Shetland, have had considerable experience in two world wars of having to cope with massive temporary influxes of population; mostly of men separated from their families, as is common in construction camps. In particular, the sheltered waters of Scapa Flow, which now attract the oil-related developments, served as a major naval base. This base was closed finally as recently as 1957. Past experience of the effects on the islands' economy of such sudden upsurges and withdrawals of population and spending power, and of the dereliction following the abandonment of sites, made Orkney County Council aware from an early stage of the need to retain control both of developments and of developers. The Council's first steps, in the autumn of 1972, were to draw up a list of sites likely to be attractive to the offshore industry, with a view to taking prior options on them itself; it also listed areas which should be preserved and protected from development. By early 1973 the Council had decided to promote a Provisional Order giving it powers to acquire land and exercise harbour jurisdiction over selected parts of its coastline; also to participate in the equity of companies proposing to introduce oil-related developments to these areas.

Fig. 6.4 shows the seaward limits of the areas of Scapa Flow and Wide Firth/Shapinsay Sound over which the County Council and its successors have been established as the harbour authority by the Act which the Council successfully promoted in 1974.[8] The inner limits are along the lines of low water fringing the shores within the harbours, and the Council have also been given powers of compulsory purchase over adjacent land areas, should they judge this to be advantageous if and when the land concerned becomes subject to oil developments. In accordance with their policy, the Council purchased the site of the Flotta terminal from the Occidental Group and then leased it back to them. They have made it mandatory on the company to obtain their approval at each stage of the development. Unlike Shetland County Council, however, the Orkney Council have not so far entered into joint investment arrangements with any oil-related company in the islands.

3. Caithness

Caithness has locational advantages for oil-related developments closely similar to those of Orkney but suffers from a lack of sheltered natural harbours. To date this disability has offset its advantages of being the most northerly mainland area of the UK and its established road, rail and airport facilities. With male unemployment running at over 8% and a substantial part of its established employment tied to the atomic power and research establishments at Dounreay, there is good reason for the local authority to try to attract oil-related employment to the area. To date, however, the only firm result from a considerable number of proposals which have been put forward is a decision to base at Wick the pilots who will guide the tankers into Flotta across the Pentland Firth. The biggest disappointment locally (though viewed quite differently by the environmental conservation lobby) has been the withdrawal of a scheme put forward by Chicago Bridge Constructors Ltd. to build steel and concrete production platforms at Dunnet Bay (Fig. 6.4). First mooted in September 1972, this scheme was projected to employ about 750 men on 42 ha (103 ac) of foreshore and levelled sand-dunes. Multiple objections to the use of such a high amenity area for industry led the Secretary of State to institute a public inquiry into the project in April 1973. By mid-October, when his decision to permit the development was announced, the Chicago Bridge Company had already obtained full permission for an alternative site in County Mayo, and in January 1974 they announced their abandonment of the Caithness site in favour of the Irish one.[9]

In the middle of 1973 two firms were also interested in developing Wick harbour as a rig-servicing base. The large breakwaters and additional quay space being suggested for one scheme would have cost some £10 million, and negotiations apparently deadlocked on the question of which body should, or could, be responsible for assembling finance on this scale.[10]

Scrabster Harbour Trust produced plans for a more modest £1 million service base in April 1974, as an extension for their newly completed roll on/roll off facility for the Orkney service. More recently feasibility surveys of both Wick and Scrabster harbours have been commissioned by Caithness County Council, but no specific proposals have been made yet for their development.

With a view to trying to determine more exactly what development opportunities are open to them, the County Council are also seeking advice from consultants on the possible value of Dunnet Bay and Sinclair's Bay as sites for pipeline landfalls, tanker terminals or offshore fabrication yards. Despite discouraging results, Caithness remains concerned to attract oil-related employment, but is over-shadowed by existing developments in areas further south. Short of improved harbour facilities being provided on a speculative basis, it seems unlikely that oil-related business will be attracted into the area unless or until the need for oil and gas pipeline landfalls exceeds the capacity of the sites already selected. Even in this respect the Inner Moray Firth has been considered a more attractive possibility by the Scottish Development Department's consultative paper on pipeline landfalls.[11]

4. The Cromarty and Moray Firths

The 100 km (62 ml) strip of land which extends roughly along the line of the present A9 and A96 between Tain and Nairn has attracted a remarkably large share of the new employment generated in Scotland by oil-related development (Table 5.6). Visually its impact is most apparent at either end of this route where massive engineering yards and large numbers of new houses attract the attention.

Like many parts of the Highlands, and for much the same reasons although here generally less pronounced, the area had for long suffered high unemployment and a consequent decline in population. This situation was most marked to the north and west of Inverness. By the early 1960s a number of factors had combined to worsen the unemployment situation. The phasing out of the former naval base at Invergordon, increasing mechanisation in agriculture and forestry, and the completion of hydro-electric schemes in the Conon and Beauly basins, had created a solid core of unemployment which was only partially relieved by seasonal tourism. Again, the effects were felt most strongly in Easter Ross and the Black Isle.

Following the setting up of the Highlands and Islands Development Board in 1965, the area of Easter Ross and the Inner Moray Firth was zoned as a major growth point. This strategy was to be achieved by attracting heavy industry to offset unemployment in order to arrest population decline. A detailed study commissioned by the Board and published in 1968 advocated a substantial increase in industrial development and population and listed some twenty communities as possible growth points.[12] Much emphasis was laid on the attractions of the area to industrialists. These included abundant flat land and water supplies, the deep water anchorages of the Cromarty Firth, a readily available supply of labour, and a combination of natural beauty and ready accessibility to outdoor and recreational pursuits. Meanwhile, with Government assistance, the Board had been successful in inducing the British Aluminium Company to establish a smelter at Invergordon. Work on this began in 1969 and helped to encourage several other companies, mainly in the petro-chemical field, to take up options on land nearby.

However, by the autumn of 1971 with construction work on the smelter nearly complete and a cut back in the company's production plans following a world recession in the aluminium industry, unemployment had once more become a serious problem, particularly in Easter Ross, and many of the new houses built in Alness remained unoccupied. It is against this background that the subsequent development of oil-related industries must be examined. Their arrival has so accelerated the development of an urban industrial complex that the Highlands and Islands Development Board's short term plans for the area have been more than fulfilled.

The deep sheltered waters of the Cromarty Firth and adjacent flat easily excavable land are key factors in the siting of rig and platform construction yards. The biggest single development has been by Highlands Fabricators Ltd (Brown & Root—Wimpey) on a 304 ha (750 ac) site at the seaward end of Nigg Bay. Work began in February 1972 on excavating a massive graving dock, 308 m x 183 m and 15 m deep (1 000 x 600 x 50 ft), and the first order for a steel platform jacket for BP's Forties Field was completed and floated out in August 1974. The jacket, code-named 'Highland I', weighed 23 000 tons and was the largest built in the world up to that time. Work also began in 1974 on a second jacket for the same field, while in November the company was successful in gaining a steel platform order for the Ninian Field.

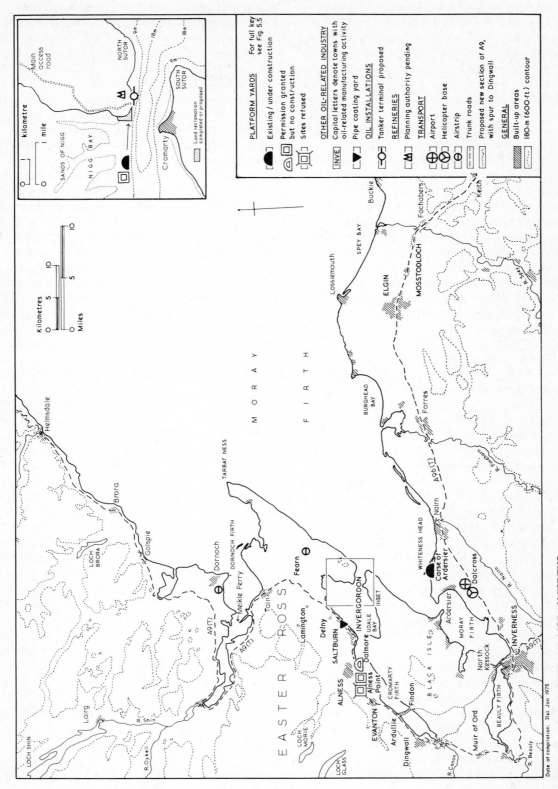

FIG 6.5   CROMARTY AND MORAY FIRTHS

Date of compilation: 31st Jan 1975

Planning permission was also obtained in 1974 to excavate a second and larger dock on a 28 ha (70 ac) site on land which will be reclaimed immediately to the west of the present dock. Work on this dock is expected to begin in 1975 and when complete it will measure 457 m x 457 m and 22 m deep (1 500 x 1 500 x 73 ft). The company intend using the second dock for building either concrete or concrete and steel hybrid platforms, and are presently engaged in preparing designs suited to the depths available at this site.

To meet their requirements the company operate a steel rolling mill with a rolling capacity of 50 000 tons per annum. Since the value of rolled steel is several times that of the original material, the use of spare capacity to process steel for other companies has proved highly profitable. The largest single order to date, 10 000 tons of rolled steel for McDermott's yard at Adersier, was received in October 1974.

A self-contained training school, opened at the yard in April 1972, has prepared nearly 1 600 men for welding and other fabricating jobs. Being recruited mainly from areas in the north of Scotland most had had no industrial experience. To establish a pool of highly trained engineers a three-year apprenticeship scheme for school leavers is also in operation, the first six months of the course being taught at Inverness Technical College.

The work force at Highlands Fabricators numbered about 2 500 at the beginning of 1975 with hopes of boosting this figure to about 3 200 by June. Although preference is given to recruiting labour in the north the present employment position in Scotland is such that most of the additional labour is likely to come from the Central Belt.

On the Cromarty Firth other fabrication yards have been proposed at Alness and Dalmore. Two of the projects, namely those of Taywood Seltrust Offshore and Mid-Continent Supply at Alness Point (concrete platforms) and Dalmore (steel platforms) respectively, received planning permission in October 1972. At present both sites remain undeveloped. A third application proposed by Costain Civil Engineering Ltd for concrete platform construction at Alness Point was rejected by the Secretary of State in December 1972.

Of the two successful applications, prospects for development seem most likely on the Taywood Seltrust site. However, this depends on the receipt of a suitable order. The same company were unsuccessful in bids to construct (on Condeep platforms on the West Highland coast (p. 106), but have recently been given permission to develop a site on the Firth of Clyde (p. 105). This may make them less interested in utilising the still shallower water site at Alness Point. Also, development of the 75 ha (184 ac) site will necessitate dredging a deep water channel to link it with the deeper water in the Firth of Invergordon.

Across the Moray Firth, J. Ray McDermott (Scotland) Ltd have developed a yard on a 333 ha (823 ac) site on the Carse of Ardersier. The company have already completed three small platforms for Phillips Ekofisk Field, the most recent (August 1974) being a 3 000 ton steel jacket for a pumping station midway along the pipeline to Teesside. Other orders have included decks and bridges for the Dutch Placid Group. At present the company have two large steel platforms under construction. The first, for the Piper Field, is now due for completion in the autumn of 1975 and the second, for the Heather Field, is due to be floated out in 1977 (Table 5.5). To provide access to the open sea a deep water channel has been dredged past the sandspit of Whiteness Head. The past year (1974) has been generally a year of consolidation for the company. Foreign labour (chiefly American) has been largely phased out, preference being given to the recruitment of local labour which is trained on the site. At time of writing the work force numbered about 800.

The only other major firm fully engaged in the offshore engineering industry is M-K Shand's pipe coating plant at Saltburn, Invergordon, although a vessel presently being fitted out as a floating engineering workshop is expected to be moored in the Cromarty Firth from March 1975. A contract for the Forties pipeline was completed by M-K Shand in the summer of 1973 and the firm is currently engaged on a similar contract, scheduled for completion in the summer of 1975, for the Frigg pipeline. The capacity of the plant was considerably boosted in the autumn of 1974 following the installation of machinery capable of producing 24 m (80 ft) lengths of coated pipe. When in full production the company employ 600 men.

The floating engineering workshop referred to above, which will be moored in the Cromarty Firth about midway between Invergordon and Cromarty, is a new concept in the repair and servicing of oil rigs. When fully operational the vessel will have a work force of about 250 and will give Scotland a stake in the important work of overhauling semi-submersible rigs, a job which could not be readily undertaken in this country previously because of the lack of shore facilities close to deep water. Further, it is hoped that the vessel will be able to put to sea in summer and carry out repairs to rigs on drilling station.

The largest investment proposed and the most controversial so far is the plan for a £150 million oil refinery at Nigg (Fig. 6.5). This plan first submitted in December 1973 by the Cromarty Petroleum Company, a subsidiary of National Bulk Carriers of New York, envisages the construction of a refinery capable of handling 200 000 barrels of crude oil daily. It is projected that about 50% of the crude oil will come from North Sea fields (the rest will be imported from the Persian Gulf) although if necessary the refinery will be able to operate entirely on North Sea oil. The main market for refinery products will be the Eastern United States. Having first had their plan rejected by Ross and Cromarty County Council in June 1974 and then later accepted for reconsideration, the company subsequently submitted an amended plan which received the Council's planning permission in principle in October 1974. However, because of the large number of objections to the scheme, the Secretary of State has been obliged to hold a public inquiry which is due to begin in Dingwall in February 1975.

The Cromarty Firth Petroleum Company have acquired 336 ha (830 ac) of land to the west of the North Sutor about one-third of which is currently zoned for industry. It is planned to site the refinery on the land nearest to the south shore with storage behind. To reduce the visual impact some 80 per cent of all storage will be underground thereby allowing the greater part of this area to continue as farmland. A marine terminal will be built with berths for 250 000 tonne tankers importing crude oil from the Persian Gulf, and inshore berths for tankers up to 80 000 tonnes to provide a shuttle service between some of the smaller North Sea fields and Nigg. When operational the refinery is expected to employ 350–400.

The realisation of the deep water facilities of the Cromarty Firth by oil-related industries, both operational and proposed, has made necessary the creation of a Harbour Authority to ensure that the waters of the Firth are used in the interests of the Scottish economy in general and the Highlands in particular, and that this is done in a way consistent with safeguarding the environment. The Cromarty Firth Harbour Authority, formed on 1st January 1974, is responsible for virtually the whole of the Firth from Findon and Ardullie to the open sea. They have power to promote the reclamation of land for development, to carry out dredging as required, to improve piers and harbours and to provide new navigational aids for shipping. Powers of reclamation, however, can only be exercised after consultation with the Countryside Commission for Scotland and the National Environment Research Council, and the Board are expressly prohibited from reclaiming any part of Udale Bay or the greater part of Nigg Bay,

both of which are widely used by migratory birds. At present (January 1975) the Harbour Authority have plans for a new pier and warehouse complex at Invergordon but both projects have encountered opposition from local people.

In addition to the industrial developments already discussed, both Inverness and Ross and Cromarty county councils have zoned land for industrial estates on a number of sites. Ownership of such land is divided between the respective county councils, the Highlands and Islands Development Board and a number of private companies and individuals. Between Ardersier and Tain some 1 134 ha (2 800 ac) have been zoned for industrial estates, the two largest blocks being at Invergordon/Delny–810 ha (2 000 ac)–and Evanton–243 ha (600 ac). On the latter estate Highland Deephaven Ltd have submitted plans to develop an industrial and marine complex of 176 ha (435 ac). Although there are other small areas zoned for industry at Alness, Dingwall, Muir of Ord and Ardersier, the remaining developments to date are in the Longman Estate in Inverness and on a site adjacent to Dalcross Airport. Both are well situated for air, road and rail transport and have ready access to the largest concentration of labour in the region.

As in north-east Scotland a number of established firms have become increasingly involved in serving the oil industry. In the area of the Cromarty and Moray Firths the scale of offshore engineering has provided opportunities for sub-contract work and the supply of machinery and tools, while steel stockholders have been attracted to establish depots. Essentially most developments are centred in Evanton/Alness, Inverness and Elgin. The location of the last mentioned, roughly midway between the Cromarty Firth/Ardersier platform yards and the service and administrative centres in the Aberdeen/Peterhead area, has encouraged two local engineering firms to become firmly committed to the oil industry and was the deciding factor in locating the steel stockyard which is presently under construction in Mosstodloch. Expanding business has encouraged one of the engineering firms to open a branch in Alness. In nearby Evanton an increasing concentration of engineering wholesalers, plant hirers and other suppliers is developing to service the fabrication yards, while in Inverness a firm which was founded one hundred years ago to manufacture agricultural implements and later branched out into railway and marine engineering are now leading specialists in all types of welding and associated techniques used in offshore fabrication work.

Urban growth, evident in the form of new housing, is particularly apparent along the north shore of the Cromarty Firth and notably in Alness, where an Aberdeen building firm recently completed their one thousandth house in the town for the County Council. Despite this there remains a severe housing shortage in Easter Ross which in practical terms means that many workers employed at the Nigg yard and elsewhere are forced to travel long distances daily to and from work, often as much as 64 km (40 ml) in either direction. To overcome the shortage of accommodation in the area in 1973–74, Highlands Fabricators had two liners moored alongside Nigg Ferry to house workers, and more recently have invested £1 million in a camp to provide accommodation for over 400 men. This is due for completion in February 1975 and a second is planned.

The housing situation has been seriously aggravated by an acute shortage of labour in the building industry. To help offset the problem, encouragement has been given by Ross-shire County Council for the erection of timber houses, mostly imported from Scandinavia. In Inverness and to the east although the situation is less critical, demand for housing both old and new remains brisk.

Population and industrial growth on the scales outlined have necessitated improvements to the existing communications network. The most notable example is the plan to re-route the A9 across the Black Isle thus greatly shortening travelling time between Easter Ross and Inverness and the south. Work is already under way on this project and if tentative plans to bridge the Dornoch Firth at Meikle Ferry are also followed through road communications along the whole of the east coast north of Inverness will be vastly improved. Rail communication has also been improved. The pipe coating plant at Saltburn is served by a specially laid spur, Alness station has been reopened to passenger traffic, and British Rail is considering ways of increasing the carrying capacity and cutting travel time on the line between Inverness and Perth. In 1974 work was carried out on extending and resurfacing the main runway at Dalcross Airport. Plans for jet services are under consideration but are unlikely to be implemented until existing landing and passenger services are improved.

The concentration of oil-related employment into a few large industrial units has hindered the growth of a balanced industrial structure in this area, particularly round the Cromarty Firth. The pressures on housing and labour have tended to frighten off smaller firms which might otherwise have set up there. This narrow industrial base is in marked contrast to the much greater diversity of employment characterised by the oil industry in Aberdeen, the next area of study.

5. The North East and Tayside

In reviews of the oil industry this area is commonly treated as two regions, the North East and Tayside. Its leading cities, Aberdeen and Dundee, are to a degree rivals, and following the reorganisation of local government in May 1975 will be capitals of the Grampian and Tayside regions respectively. Industrial promotion has been handled independently by NESDA (the North East Scotland Development Authority) and the Tayside Development Authority. From May 1975 these will become the development departments of the two regional administrations. From many points of view it would seem logical to consider the areas separately, but oil-related developments from Kinnaird's Head to Fife Ness are so interconnected that it makes sense to review them as a whole.

The whole industry has grown outward from, and maintains its organisational focus in, Aberdeen. New port developments at Peterhead, Montrose and Dundee were undertaken primarily because Aberdeen's harbour could not be extended or refitted rapidly enough to cater for the explosive growth in rig-servicing traffic. All three remain in part tributary to Aberdeen, although Peterhead has also developed an independent role in servicing pipelaying operations and Dundee, with marine bases for two major oil companies, has become a competitor for rig and platform servicing.

With the technical organisation and control of exploration and development too strongly entrenched in Aberdeen for Dundee to challenge this role, Dundee's greatest hope for expansion in the oil industry, apart from oil field servicing, would seem to lie in engineering. The city is also making a strong bid to attract the headquarters of the proposed British National Oil Corporation. In comparison with Aberdeen, Tayside has less strongly established service industries and limited air communications, but its more strongly developed manufacturing structure, higher unemployment rates, greater availability of housing, and marginally better land communications provide a basis to be exploited. The oil industry on Tayside has been slow to establish a functional identity separate from that of the North East, but attraction to the area of a few more large concerns could change this situation.

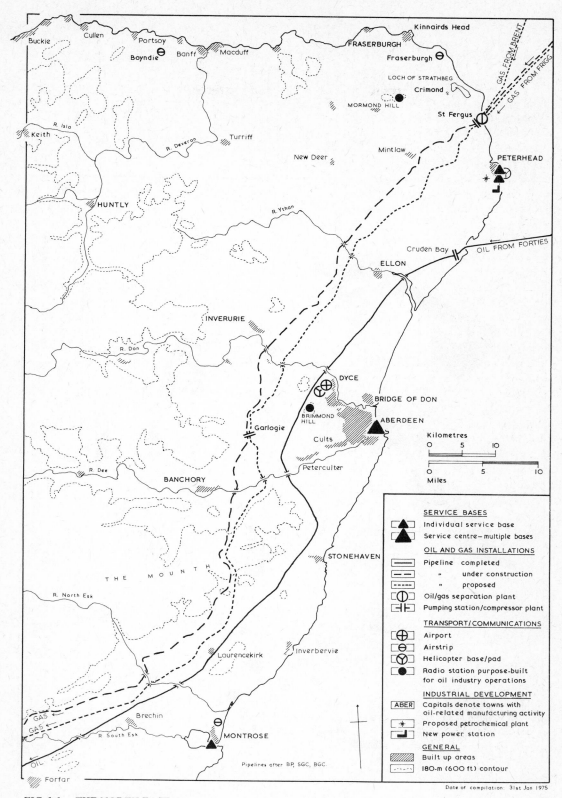

FIG 6.6   THE NORTH EAST

Within the map:

Buckie · Cullen · Portsoy · Banff · Macduff · FRASERBURGH · Kinnairds Head
Boyndie · Fraserburgh
LOCH OF STRATHBEG
Crimond
MORMOND HILL
St Fergus
GAS FROM BRENT
GAS FROM FRIGG
R. Isla · Keith · R. Deveron · Turriff · New Deer · Mintlaw · PETERHEAD
HUNTLY · R. Ythan
Cruden Bay · OIL FROM FORTIES
ELLON
INVERURIE
R. Don
DYCE
BRIDGE OF DON
BRIMMOND HILL · ABERDEEN
Garlogie · Cults
Peterculter
R. Dee · BANCHORY
T H E   M O U N T H
STONEHAVEN
R. North Esk
GAS · GAS · Brechin
R. South Esk
OIL · Laurencekirk · Inverbervie
MONTROSE
Forfar

Kilometres
0   5   10
0   5   10
Miles

Pipelines after BP, SGC, BGC.

SERVICE BASES
▲ Individual service base
▲ Service centre – multiple bases

OIL AND GAS INSTALLATIONS
Pipeline completed
 "   under construction
 "   proposed
Oil/gas separation plant
Pumping station/compressor plant

TRANSPORT/COMMUNICATIONS
Airport
Airstrip
Helicopter base/pad
Radio station purpose-built for oil industry operations

INDUSTRIAL DEVELOPMENT
ABER Capitals denote towns with oil-related manufacturing activity
* Proposed petrochemical plant
New power station

GENERAL
Built up areas
180-m (600 ft) contour

Date of compilation: 31st Jan 1975

81

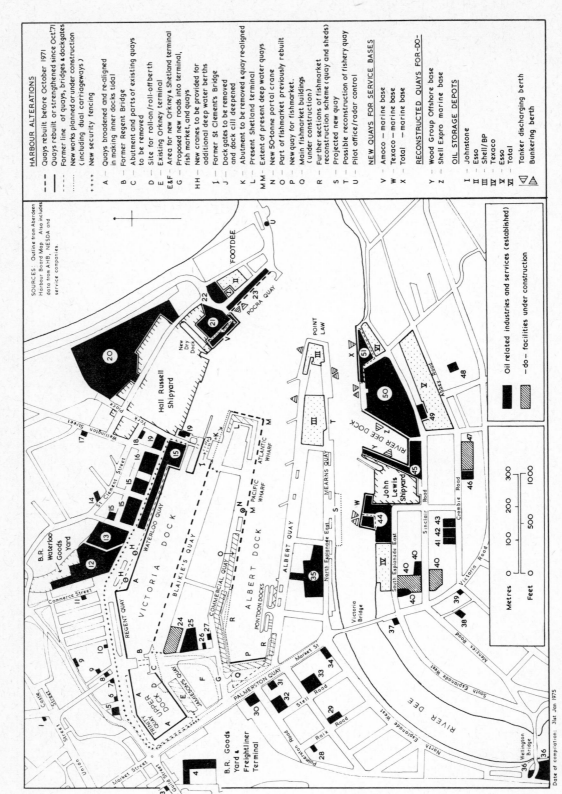

FIG 6.7   ABERDEEN HARBOUR

HARBOUR ALTERATIONS

Quays rebuilt before October 1971

Quays rebuilt or strengthened since Oct.'71

Former line of quays, bridges & dockgates

New works planned or under construction
(including dual carriageways)

New security fencing

A — Quays broadened and re-aligned
in making inner docks tidal

B — Former Regent Bridge

C — Abutment and parts of existing quays
to be removed

D — Site for roll-on/roll-offberth

E — Existing Orkney terminal

E&F — Area for new Orkney & Shetland terminal

G — Proposed new roads into terminal,
fish market, and quays

H H — New cranes to be provided for
additional deep water berths

J — Former St Clement's Bridge

J — Dock gates to be removed
and dock cill deepened

K — Abutment to be removed & quay re-aligned

L — Present Shetland terminal

M M — Extent of present deep water quays

N — New 50-tonne portal crane

O — Part of fishmarket previously rebuilt

P — New quay for fishmarket.

Q — Main fishmarket buildings
(under construction)

R — Further sections of fishmarket
reconstruction scheme (quay and sheds)

S — Projected new quay

T — Possible reconstruction of fishery quay

U — Pilot office/radar control

NEW QUAYS FOR SERVICE BASES

V — Amoco — marine base

W — Texaco — marine base

X — Total — marine base

RECONSTRUCTED QUAYS FOR—do—

Y — Wood Group Offshore base

Z — Shell Expro marine base

OIL STORAGE DEPOTS

I — Johnstone

II — Esso

III — Shell/BP

IV — Texaco

V — Esso

VI — Total

▽△ — Tanker discharging berth

△ — Bunkering berth

SOURCES: Outline from Aberdeen
Harbour Board Map. Also includes
data from AHB, NESDA and
service companies.

Date of compilation: 31st Jan 1975

■ Oil related industries and services (established)

▨ —do— facilities under construction

Metres 0   100   200   300

Feet 0   500   1000

From the earliest stages of exploration in the mid-1960s Aberdeen has played an important part in serving the oil industry in the northern North Sea. Since 1970 the scale and scope of operations based in the city have grown so spectacularly that it has not only become the oil 'capital' of Scotland but has drawn to it the largest single concentration of specialist offshore companies and personnel in Western Europe. Nearly all the companies prospecting in Scottish waters, most of the drilling contractors, manufacturers of drilling equipment and chemicals, and specialist firms providing technical services and expertise have premises in Aberdeen. In addition, six of the major oil companies have chosen the city as administrative headquarters for their whole UK programme of North Sea exploration and development, and Aberdeen harbour maintains its clear lead as the principal UK offshore supply and service point.

The city owes its pre-eminence in this multi-million pound industry to four main factors—its port, its airport, its well-established service industries and its enterprise. Being first in the field as a supply centre has given it a lead which other places find difficult to challenge, and a great many agencies locally have worked diligently and consistently to develop and maintain that lead. Trade missions have been organised to the USA, space regularly taken at major exhibitions and conferences at home and abroad, and publicity and information material produced and distributed on a large scale. Research and teaching in a variety of subjects of interest to the oil industry has been actively developed in the local university and colleges. Possibly most effective of all, large numbers of people have been brought to the city to see for themselves the scale of development and the advantages of becoming involved in it. This activity goes on all the time, but reached a peak in the highly successful Offshore Scotland trade exhibition held in March 1973. It should be substantially surpassed by the more ambitious Offshore Europe '75 Exhibition and Conference which is expected to bring over 20 000 visitors to Aberdeen in September 1975.

It would be wrong, of course, to suggest that the oil industry has found the city's facilities in all ways ideal, or that the pressures of development have not brought difficulties as well as advantages to the area. Until

---

FIGURE 6.7 KEY TO NUMBERED FEATURES*

1. Ocean Systems UK
2. British Gas Corporation
3. Motherwell Bridge/Strongwork Diving
4. ASCO Guild Base**
5. Aberdeen Harbour Board
6. Sea Oil Services/Bugge Supply Ships
7. John Cook & Son
8. Offshore Supply Association/Smith Lloyd/Marine Safeway/Nauticom
9. Shore Porters Society**
10. Oakes Welding
11. Welsh Simes and Salvesen
12. British Steel Corporation
13. Maxwell Properties**
14. British Ropes
15. Seaforth Maritime—Seaforth Seabase**
16. Sea Oil Services**
17. Incon Offshore
18. North Sea Gas Services
19. Hudsons Offshore**
20. Hudsons Nordcentre
21. Amoco marine base
22. Dresser Oilfield Products
23. Silos (various mud companies)
24. Christian Salvesen (office block and warehouses under construction)
25. ASCO Regent Base**
26. Marconi International Marine
27. Clifford Oilfield Services
28. John Brown and Sons
29. John Wood Group Offshore
30. Cosalt
31. J. Marr
32. ASCO Palmerston Base**
33. Total Oil Marine
34. Lloyds British Testing
35. John Wood Group Oil Services (Aberdeen)/Wood-Weir Engineering/Weir-Houston/Wood and Davidson
36. Arbarthorpe Oilfield Services**
37. Climax Engineering Services (Offshore)
38. Wilson Walton International
39. Supply Ship Services
40. Maxwell Properties**
41. Milchem
42. Oilfield Manufacturing and Services
43. Zapata Offshore Services
44. Texaco marine base
45. John Wood Group Offshore marine base
46. Hunting Oilfield Services
47. Sunley Investment Trust (office block and warehouses under construction)
48. East Anglia Electrical Group/Lion Oil Tools
49. Stormdrill
50. Shell Expro marine base
51. Total Oil Marine marine base

* Firms are shown by name (omitting Co., Ltd, Inc., etc). The nature of their activities is shown by symbols on Fig. 6.8.
** Details of the number and types of firms located at these service bases or property developments appear in the key to Fig. 6.8.

83

recently port improvements were not able to keep pace with demand. Improvements to Dyce Airport have been introduced even more slowly in spite of huge increases in traffic. In the initial stage of explosive growth, space for offices and warehouses was largely found by determined innovation in very restricted circumstances. Firms have been able only gradually to find or build premises adequate to their needs, and storage at all close to the harbour will remain in restricted supply. Steeply inflated house prices have hit the purses of incomers as well as of local people.

Why then did the industry choose to come to Aberdeen in the first place? Why was this particular point along the coast singled out? From the beginning the prime focus of activities has been the harbour. When oil prospecting started in Scottish waters in the mid-1960s only four ports could provide the 24-hour, all-year-round availability needed by survey and supply vessels— Grangemouth, Leith, Dundee and Aberdeen. Of these, only Leith and Aberdeen had air services adequate to the needs of the industry (Fig. 5.3) and, of all four, Aberdeen had the best accessibility to the sea areas being searched (Fig. 5.2) and was very seldom closed by bad weather. As a major fishing port, holiday resort and regional service centre, the city already had many of the facilities which would be needed to support the exploration work (Fig. 5.7) and, indeed, in the early years this activity was accommodated almost unnoticed by the population at large. As recently as the summer of 1969 there were no indications that the search in the northern North Sea was on the point of success. In July 1969 the North East's daily newspaper, *The Press and Journal,* which has throughout given exemplary coverage to oil-related news, devoted only 33 words of a sixteen-page 'Review of the City of Aberdeen' to the search for oil and gas. These were mainly a warning against placing too much hope in its success.

In the later 1960s the main economic concern in the North East was its continuing loss of population, a net loss of around 4 500 per year, in the face of inadequate job opportunities. As part of the regional effort to attract new employment Aberdeen Town Council was advertising advance factories and 99-year leases of land on its industrial estates, at very low rents.[13] Aberdeen Harbour Board, already spending considerable sums to retain traffic through improvements to its commercial quays, could see no immediate prospect of raising the £500 000 needed to repair a section of an old fishery quay which had collapsed. Aberdeen's oil boom really started in the autumn of 1970, following BP's discovery of the Forties Field. Within a year pressure of demand on the port for

quay space and accessible storage had so built up that BP transferred its marine base to Dundee and the first steps had been taken to create a service base at Peterhead.

Figs. 6.7 and 6.8 show that the oil industry in Aberdeen has had to be fitted into a harbour which is the third largest fishing port in the UK, an active commercial port, and the site of two shipbuilding yards. The Victoria and Upper docks were enclosed basins, with movement into and out of them restricted to about four hours at each high tide, so they were of limited use to vessels servicing rigs (p. 48). The larger freighters using the harbour naturally had priority at the deep water wharves, and other tidal areas of the harbour were largely given over to the trawler and inshore fleets.

Until additional berthage could be constructed, the pressure of handling the service vessels fell mainly on Pocra Quay and Maitland's Quay East (Fig. 6.7, Z & 23), with Point Law and the deep water quays (M) being used whenever possible. By good fortune Pocra Quay had been decked in concrete a few years earlier, and so was able to bear heavy silos of drilling muds and cement in addition to the well casing pipes, rig anchors and chains and containers of many types, which had to be loaded on to the service ships. Fuel and water, needed by the rigs as well as the ships, was available from oil bunkering depots and pipelines already installed to cater for the fishing fleet.

It was obvious from an early stage that these facilities were altogether inadequate for the service vessel traffic likely to use the port. Substantial increases in the tidal quayage were essential. Among ambitious extension schemes considered were the building of a new dock in Nigg Bay (Fig. 6.8) and the creation of a massive new basin to the north of the present North Pier. Construction of either, even if practical, would have been too slow to meet the immediate pressure of demand, so attention was concentrated on remodelling the space within the existing harbour. Works in hand to date represent a total capital investment of £12 million.

The main scheme, adopted by the Harbour Board in October 1971, involved major reconstruction of quays in the Victoria and Upper docks and removal of their dock gates to make the entire harbour tidal. Estimated initially at £1.3 million it was scheduled for completion by May 1974, but delays arising from a variety of causes have raised the cost to over £3 million and put back the target date for completion to the summer of 1975. This major improvement has meanwhile been supplemented by the construction of five further stretches of new quay by private interests, four of these to create private marine bases for individual oil companies and one a multi-user facility (Fig. 6.7).

Reconstruction work undertaken by the Harbour Board has included realigning quays to provide longer clear stretches of working area and to accommodate construction of two new dual-carriageway roads alongside the Upper Dock. The inner docks have been dredged to provide a minimum of 6 m (20 feet) throughout at low water Spring tides (LWS), with up to 10 m (32 feet) alongside Waterloo Quay. Additionally, plans have been approved by the Government for deepening the entrance to the inner docks and removing the abutment there at the cost of a further £1¼ million. This will allow passage at all states of the tide to the increasingly large oil service vessels now coming into service, and also to the new Shetland ferry which will operate by the middle of 1976.[14]

A roll-on/roll-off terminal for this new vessel is to be provided at the upper end of Victoria Dock, as shown on Fig 6.7, the facility being also available to other vessels, including oil-support ships. Its site has in part been created by removal of the former Regent Bridge which, with another bridge at the former lock gates, has been removed to give shipping greater freedom of movement within the harbour. Also new, at the deep water wharves, is a 50-ton portal crane introduced to help handle the increasingly large commercial traffic of the port, including regular services by ships bringing oilfield equipment from the US Gulf ports. Two cranes will also be provided this year at Waterloo Quay to service the new deep water berths there, which will supplement the Atlantic and Pacific Wharves.

Handling of the greatly increased number of movements of vessels, both within and entering the port, has made necessary the installation of a system of radar surveillance and the recruitment of a team of navigational control officers to organise the harbour traffic on a 24 hours-a-day basis.[15] Yet, surprisingly, there is still an active salmon netting fishery operating from the south bank of the River Dee within the harbour confines. Rents from this source are still a consideration in the revenues of the Harbour Board, and schemes for the construction of further quays for oil traffic along the River Dee section of the harbour must be weighed against the elimination of the fishery.

Four of the privately developed marine bases in the harbour are already along the banks of the Dee, two developed from waterfront property which the firms concerned previously used for other purposes and two by displacing fishing interests or housing. Texaco (44) have built an extension on to their existing oil importing and bunkering depot. The Wood Group, a local fishery

and engineering combine which has rapidly diversified into the oil servicing industry, has created a 'one stop shopping' base (p. 49) on quays and land held by its shipbuilding subsidiary (45). Shell Expro has rebuilt the remaining, east, side of the River Dee Dock and part of the former village of Old Torry (pp. 113) to form its marine base, while Total has filled in Torry Harbour, where the smallest inshore fishing boats used to be moored, for its share of the waterfront.

The extreme importance which such firms attach to having a degree of independence in the control of loading and turn-round of ships serving the rigs which operate for them is evidenced by the way they are prepared to spend money in getting control of even such small units of waterfront land as Total gained in this case. The rather larger, but still limited, frontage (V) which Amoco has developed alongside the Hall-Russell shipyard had at one stage been considered by the Harbour Board for reconstruction, but the project was set aside because the area to be gained would be so small.

In all cases these firms have had to find additional land elsewhere in the city on which to establish their main storage depots and offices, and the same is true of the remaining privately operated marine base developed by Seaforth Maritime (15).[16] This company, like the Wood Group, operates a base which is open to custom, on contract or otherwise, of any ships which need its services. It differs from other private bases in making use of part of the new quays constructed by the Harbour Board, as will also Salvesen's 11-storey office and oil-support facility which is under construction at the opposite corner of Victoria Dock (24).[17]

In February 1974 Seaforth Maritime obtained exclusive use of 200 m of the reconstructed Waterloo Quay, much of this immediately in front of the large warehouse and office block they were developing from a former grain mill. This provides three berths, and at a later stage may be extended on to the 'island' at the mouth of Victoria Dock to provide a fourth berth. 'One-stop shopping' (p. 49) will be provided when a pipeline is laid to convey fuel oil directly to the berths from the storage depot behind Pocra Quay. All their other liquid and powder requirements are already pumped to service vessels through ducts specially laid in the quay.

The 'one-stop' principle, has been developed as fully as possible at all the private bases in the harbour, but vessels using the reconstructed 'public-user' quays at Aberdeen will often make two stops, calling at Pocra Quay for mud, water and oil, and having other supplies brought to them by road transport at another quay. Time penalties involved in the additional ship movements should be small, and the harbour has the basic advantage of being able to use most quays flexibly for both oil-support and general cargo traffic—a necessary precaution for the continued commercial life of the port.

In this respect long established port authorities such as Aberdeen and Dundee have much more reason to think ahead to the days when oil support traffic will decline than do those concerns which have come into being simply to supply oil base facilities. Likewise, the Harbour Board have a major responsibility to the fishing industry. A £2 million reconstruction scheme of the quays and fishmarket round the head of the Albert Basin has been pushed ahead simultaneously with other harbour improvements, and fishing interests were allocated some of the extremely scarce land between the quays which was vacated by an animal feedstuff firm during 1974.

The oil industry has put tremendous pressure on space close to the harbour and the Harbour Board, as owners of much of this land, has encouraged tenants for whom a dockside location was not essential to move elsewhere. Others, such as the firm just mentioned, have moved out because they considered that the increasing pressure of traffic in the dock area was restricting their operations. British Rail and its associated companies, owning the second largest area close to the harbour, made available for development both land and buildings which they were no longer fully utilising. Reconstruction of the greater part of Waterloo Goods Yard as a well casing stockyard for the British Steel Corporation is a case in point. The yard could hardly be more conveniently sited for transfer of the heavy pipes to supply ships and British Rail, in addition to rent for the space, have revenue from transporting the pipes in full train loads direct from the steelworks.

Both shipyards obviously hold prime sites for port redevelopment, but they have reacted rather differently to the challenges set them by the advent of the oil industry. The Wood Group, as mentioned above, has converted part of its repair and fitting-out berths to a supply ship service base. The Hall Russell yard, in addition to undertaking engineering repair work for the industry, has added rig supply vessels to the already wide range of ships it produces, and has almost completed construction of a large dry dock which will give it the capacity to overhaul and repair vessels up to 113 m (370 ft) in length by 18 m (60 ft) in beam. Despite increasing difficulty in retaining their skilled tradesmen in competition with the high wages offered by firms more closely connected with the oil industry, the shipyards are fighting actively to realise the advantages of their position in this key harbour. The pressures of

incoming firms for both space and labour, however, remain intense.

A feature of firms in the oil search is their apparent need to locate close to one another. In exploration and development work immediate 'trouble-shooting' contacts between oil companies, their specialist suppliers, shipping agents and technical advisers are often needed to bring expertise to bear on problems as these arise. The oil industry also expects those it deals with both to advertise themselves actively and to be on immediate call when needed. As a result, in the fashion of the financial and commercial interests of the City of London, it is important for all those concerned with the oil industry to have at least an office, and preferably also stores and workshops, as close as possible to the hub of operations, in this case the harbour.

Many firms need only a room or two in which to establish a bridgehead, and most of them, unless they are engaged in the longer term work of developing an already defined discovery, want to rent space rather than be involved in purchase and maintenance of premises. Their needs have been provided for by the growth of a series of other companies which have created service bases or have set up organisations to lease space and provide transport, freight-handling and similar facilities.

These service bases and centres vary widely in size, style and organisation. Many have been produced by building suites of offices into old warehouses and factories, with the remaining storage space in them being leased in smaller units to separate tenants. When the scale and likely permanence of the oil industry made itself pressingly obvious in the early 1970s more elaborate conversions were rapidly developed, but never fast enough to meet the demand for space. With larger concrete or steel-framed buildings constructed in former yards or taking the place of demolished buildings, these newer premises are better suited to modern methods of mechanical handling. Some are close to the docks, but most developers have had to settle for sites rather further away, especially on industrial estates where completely new buildings have been constructed. (Fig. 6.8).

Table 6.1 gives an indication of the rate at which new firms directly involved in the oil industry were setting up in Aberdeen and District over a three-year period, and the numbers of people employed by them. Between October 1972 and October 1973 such firms were establishing themselves in the area at an average rate of two a week. Compared with this frantic rush the more recent phase since late 1973 has been

TABLE 6.1 ABERDEEN AND DISTRICT, GROWTH OF
ONSHORE SERVICES 1971–1974

	Number of firms involved in the offshore oil industry as a principal activity	Approximate numbers of employees involved
October 1971	56	n/a
October 1972	109	1 500
March 1973	153	n/a
June 1973	193	n/a
October 1973	217	3 000
February 1974	202	4 000
August 1974	236	5 500

Notes
i) Figures derived from various issues of *North East Scotland and the Offshore Industry*, published by NESDA.
ii) At each of these dates a further group of firms, half as large again, was supplying the oil industry as a 'partial' activity, i.e. without committing the major part of their resources to this market.
iii) 'Aberdeen and District': almost all the firms concerned were located within the area of Fig. 6.8.

one mainly of consolidation, but the numbers employed are still rising steadily.

The overall and continuing shortage of accommodation, especially of warehouse space, the system of leasing from service bases, and the rapidly changing fortunes of individual companies when new rigs become available or 'strikes' are made offshore, have combined to produce an extremely volatile and dynamic location pattern. As their operations have expanded, many firms have had to spread their activities into a number of quite widely scattered sites. Once committed to the oil industry even the oldest and most awkwardly located premises have typically remained fully occupied. When new buildings are completed for individual firms their existing offices and warehouses are taken up by neighbours who have outgrown their own space, or by incoming small enterprises which have yet to make their mark locally.

Much of the equipment concerned in oilfield exploration and operation is extremely bulky. For instance approximately 400 tonnes of steel casing in 13 m lengths and 365 tonnes of cement are typically required to line a single well. The 'string' of drill pipe at the end of which the drilling bit is rotated can be anything from 2 100m to 4 200m long (7 000–14 000 ft). It has been obvious from the start that only a small proportion of this gear could be stored at all close to the harbour so most operators have settled for having at least two, and possibly three or four sets of premises. Firms such as

Shell Expro, Amoco and Texaco, despite having their own marine bases and substantial storage areas of their own on industrial estates within the city, typically rent open storage or warehousing elsewhere from a variety of service companies.

Examination of where these and other oil and drilling firms are located, and of some of the most recently developed service centres, shows how much the expansion of the oil-associated industries in Aberdeen has depended on land made available in the industrial estates of Aberdeen city and in the adjacent county of Aberdeenshire (Fig. 6.8). It was fortunate for the industry that the Town Council in the 1960s had serviced substantial areas of these estates, only a short distance to the south of the harbour. More recently, extension of the city boundary south-eastward made possible the re-zoning for industry of further land at Altens, but there have been substantial delays in making a start to factory building there, largely because of difficulties in providing a water supply for the site.

Meanwhile, the shortage of land within the city boundary has encouraged developers to look further afield. Two service companies have developed open storage yards in Kincardineshire about seven kilometres (4½ miles) to the south of the harbour and a large number of firms have moved into estates to the north and north-west of the city in Aberdeenshire. Advantages claimed for the Kincardine sites are the lower cost of land compared with Aberdeen and their relative proximity to the harbour without the need to negotiate crowded streets through the city and northern suburbs. Developers are understood to hold options on substantial acreages of land in this area so further schemes of this kind are a possibility. Decisions will rest with the new Kincardine and Deeside District, as planning authority. This area, unlike the rest of that shown on Fig. 6.8, lies outwith the new Aberdeen City District.

Rapid growth in recent years of industrial areas to the north of the city has owed much to the fact that the Town Council's estates were too small to meet the combined demands of the oil industry and of local firms which were being displaced by redevelopment schemes. The Farburn Estate at Dyce was almost fully allocated at a single meeting of Aberdeenshire Planning Committee in 1972. The larger Bridge of Don Estate was also filled up very rapidly, and the first of two planned extensions to it is now being serviced (January 1975). Five further estates are in process of development close to Dyce Airport, three by the County Council and two by private developers, and both the Council and a private firm have further substantial areas to the west of the air-

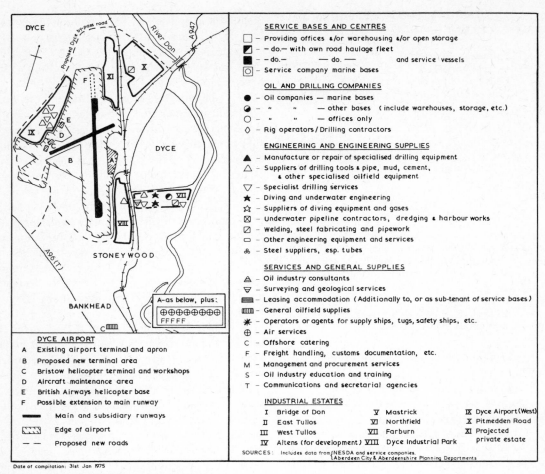

DYCE AIRPORT

A Existing airport terminal and apron
B Proposed new terminal area
C Bristow helicopter terminal and workshops
D Aircraft maintenance area
E British Airways helicopter base
F Possible extension to main runway

▬ Main and subsidiary runways
▨ Edge of airport
– – Proposed new roads

SERVICE BASES AND CENTRES

☐ – Providing offices &/or warehousing &/or open storage
◩ – do.– with own road haulage fleet
■ – do.– — do. — and service vessels
◎ – Service company marine bases

OIL AND DRILLING COMPANIES

● – Oil companies — marine bases
◑ – " " — other bases (include warehouses, storage, etc.)
○ – " " — offices only
◇ – Rig operators / Drilling contractors

ENGINEERING AND ENGINEERING SUPPLIES

▲ – Manufacture or repair of specialised drilling equipment
△ – Suppliers of drilling tools & pipe, mud, cement, & other specialised oilfield equipment
▽ – Specialist drilling services
★ – Diving and underwater engineering
☆ – Suppliers of diving equipment and gases
⊠ – Underwater pipeline contractors, dredging & harbour works
◪ – Welding, steel fabricating and pipework
▭ – Other engineering equipment and services
♣ – Steel suppliers, esp. tubes

SERVICES AND GENERAL SUPPLIES

◭ – Oil industry consultants
▽ – Surveying and geological services
▤ – Leasing accommodation (Additionally to, or as sub-tenant of service bases)
▥ – General oilfield supplies
✳ – Operators or agents for supply ships, tugs, safety ships, etc.
⊕ – Air services
C – Offshore catering
F – Freight handling, customs documentation, etc.
M – Management and procurement services
S – Oil industry education and training
T – Communications and secretarial agencies

INDUSTRIAL ESTATES

I   Bridge of Don
II East Tullos
III West Tullos
IV Altens (for development)
V   Mastrick
VI Northfield
VII Farburn
VIII Dyce Industrial Park
IX Dyce Airport (West)
X   Pitmedden Road
XI Projected private estate

A–as below, plus:
⊕⊕⊕⊕⊕⊕⊕ FFFFF

SOURCES: Includes data from NESDA and service companies.
Aberdeen City & Aberdeenshire Planning Departments

Date of compilation: 31st Jan 1975

FIG 6.8   ABERDEEN AND DISTRICT

port designated for use when these are taken up.

Two particular features of note about the industrial developments to date to the north of the city are the establishment of oil tool manufacturing by US companies at Bridge of Don and the airfield-related nature of a number of the estate tenants at Dyce. Three subsidiaries of internationally known oil tool manufacturers have set up at Bridge of Don, namely Vetco Offshore, Baker Oil Tools and Smith International. With a combined work force at present of about 250 and plans to expand in the medium term to emply 500–600, these three factories represent a considerable foothold for the area in the manufacturing side of the oil business. As at the Halliburton factory at Arbroath (p. 95), the parent companies have allocated to each factory as sales area for its products not only the North Sea market but the whole field of Europe and the Middle East, a feature of key importance for the long-term prosperity and expansion of the plants.

The increasing consolidation of industrial development at Dyce helps to emphasise how from the beginning the airport there has provided a twin and complementary focus to the harbour in the establishment and development of the oil industry at Aberdeen. Patterns of land use introduced into the Aberdeen area by oil-related business might well be compared to the ever-widening ripples which move outwards across the surface of a pond when a stone is thrown into it. Their centre is primarily the harbour, and there have been multiple sets of these ripples set up as successive leasings of exploration blocks—and finding of oil and gas in some of them—have brought increasing activity to the port. But from its inception the oil industry has also depended on the capacity of Dyce Airport to handle a wide range of short and medium range aircraft, also the helicopter services needed to provide direct connections with the rigs and lay barges.

The frequency and worldwide scale of travel

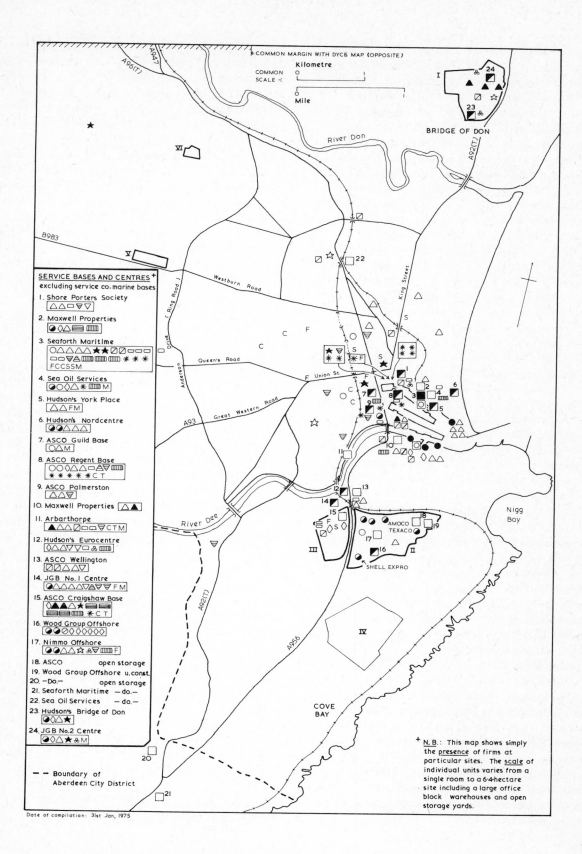

COMMON MARGIN WITH DYCE MAP (OPPOSITE)

Kilometre

COMMON
SCALE

Mile

BRIDGE OF DON

River Don

B983

SERVICE BASES AND CENTRES +
excluding service co. marine bases

1. Shore Porters Society

2. Maxwell Properties

3. Seaforth Maritime
FCCSSM

4. Sea Oil Services

5. Hudson's York Place

6. Hudson's Nordcentre

7. ASCO Guild Base

8. ASCO Regent Base

9. ASCO Palmerston

10. Maxwell Properties

11. Arbarthorpe

12. Hudson's Eurocentre

13. ASCO Wellington

14. JGB No.1 Centre

15. ASCO Craigshaw Base

16. Wood Group Offshore

17. Nimmo Offshore

18. ASCO   open storage

19. Wood Group Offshore u. const.

20. —Do.—   open storage

21. Seaforth Maritime —do.—

22. Sea Oil Services —do.—

23. Hudson's Bridge of Don

24. JGB No.2 Centre

– – Boundary of
Aberdeen City District

Date of compilation: 31st Jan, 1975

Westburn Road

Queen's Road

Union St.

Great Western Road

River Dee

King Street

Nigg
Bay

AMOCO
TEXACO

SHELL EXPRO

COVE
BAY

+ N.B.: This map shows simply
the presence of firms at
particular sites. The scale of
individual units varies from a
single room to a 6·4-hectare
site including a large office
block warehouses and open
storage yards.

89

which is a commonplace feature in the lives of oil company executives and technical specialists has brought a huge growth in the numbers of scheduled, charter and air-taxi services operating from Dyce. With the numbers of passengers rising from 135 000 in 1970 to 450 000 in 1974, the airport has severely outgrown its terminal facilities despite a series of prefabricated extensions. It has also created a steadily increasing noise problem for people living under the main flight paths on the north-west side of the city. Plans are now in hand for a completely new passenger terminal to be built by 1977 at the south-west corner of the airport, financed by a £6 million loan from the Government. It will have access from the proposed Dyce bypass road at the same side of the airfield as the two helicopter terminals. Meanwhile a further temporary building is under construction to provide separate handling of passengers on international flights, and a radar system of air traffic control is to be introduced within the next year.

Although a number of large pieces of urgently needed rig equipment have been brought direct from the US to Dyce by specially chartered air freighters, limited handling facilities for other than fairly small items have so far curtailed the growth of freight traffic through the airport. Helicopter traffic has, however, become so important that both British Airways and Bristow Helicopters have established their own maintenance bases and passenger facilities at Dyce. Subsidiary bases at Sumburgh cater for oil operations off Shetland, but with the effective range for helicopters in use being around the 320 km (200 ml) mark Dyce has been well situated to serve exploration and production areas off the east coast (Fig. 5.2).

In addition to the regular rig crews, who have to be ferried out and back on a weekly basis, there are four or five other flights to be made to each rig in the course of a week, carrying specialist personnel and urgent supplies. The need of some firms for immediate access to the heliport was a factor taken into account when allocating tenancies in the neighbouring County Council industrial estate (Fig. 6.8, IX), and a heavy dependence on the airport in general is a feature of a number of firms which have located at Dyce. Examples include oilwell testing, directional drilling, underwater engineering and similar highly specialist firms whose experts are almost constantly on the move from one major oilfield area to another. Significantly too, BP have established their UK headquarters for exploration and production at Dyce despite transferring their marine operations to Dundee in 1971 (pp. 84 and 95).

Encouraged by the likely attractiveness to oil-related firms of sites close to the airport several property development companies have bought land at Dyce for industrial estates to be operated in competition with the local authority ones. So far these private enterprise units have been slow to attract tenants but they have control of substantial areas of prime sites.[18] Although it may not bring an immediate return on the investment made, there should be good prospects of such sites being taken up in the future, because exploitation of the oil and gas in the northern North Sea has yet to reach its peak.[19] Continued development of Aberdeen and District as the supply and organising focus of the offshore industry can be expected to result, and the major share of the land zoned for industrial use is here at Dyce.

B: The Peterhead Area

Although the distances are greater, the location of East Aberdeenshire relative to the oil and gas finds off the Scottish coast closely parallels that of East Anglia to the gas fields of the southern North Sea. Aberdeen has assumed the role played in the south by Great Yarmouth, and the Peterhead area, as well as sharing in the offshore servicing, is providing landfalls for undersea pipelines similar to those of Bacton in Norfolk. A pipeline from the Forties Field to Cruden Bay was completed in 1974, and one of a pair of further large diameter pipes which are to bring gas from both the Norwegian and UK sections of the Frigg gas field is under construction to St Fergus. Laying of a third line to bring gas from the Brent Field is under consideration.

Altogether exceptional pressure has been placed upon the town and area of Peterhead in the past three years. Development of the two landfalls has coincided with the rapid construction and traffic growth of two new oil service bases at Peterhead, as well as the early stages of establishing a large new oil-burning power station just south of the town; and all of these have come on top of a period of spectacular growth of fish landings at the port. Property speculators were quickly in the field, buying options on large areas of land around the town, and Peterhead has had its full share of attention from researchers and journalists fascinated by the impact of the influx of people and wealth on the economic and moral fabric of the place.[20] The account which follows attempts simply to chart the location and growth patterns of development to date.

Factors influencing the general location of pipeline landfalls have been already considered in Chapter 5, and related planning issues are covered in Chapter 7, but it should be pointed out here that there is a very great contrast in the size and nature of the onshore developments at St Fergus and Cruden Bay. For the Forties pipeline there is only a small installation of some 8 ha (20 ac), on farmland about 1½ km south of the landfall. It has surge tanks to act as safety valves against undue build-up of pressure in the submarine pipe, and a pumping station to propel the oil onward to Grangemouth. The plant at St Fergus is much larger and more elaborate. With 200 ha (500 ac) it may ultimately become the largest gas terminal in Europe.

Only about half of the St Fergus site is to be used initially, leaving space for handling supplies from possible additional pipelines from other fields. At the seaward end of the site the gas field operators (Total Oil Marine for Frigg) will separate water and hydrocarbon liquids from the gas, which has then to be cleaned and adjusted to the correct thermal values for general consumption before being despatched south. For this, the British Gas Corporation's share at the landward end of the St Fergus site includes huge industrial jet engines, themselves fuelled by natural gas, which drive compressors to raise sufficient pressure in the outgoing pipelines. Further compressor stations covering about 24 ha (60 acres) each, which are needed every 64—80 km (40—50 ml) along the land pipelines, are shown on Figs. 6.6 and 6.11.

These maps also show that by the time the two gas pipelines from St Fergus are completed in 1976 certain sections of the available lowland routes south from the Peterhead area to the central belt of Scotland will have become quite 'crowded' with pipeline routes. For safety reasons, and to avoid putting over-large sections of individual farms out of action while the pipes are laid, the two 36″ gas pipes are not being laid in the same trench. In some cases they will lie two or more miles apart. If it becomes necessary to construct further land pipelines from St Fergus carrying gas south from other fields such as Brent, a considerably more congested pattern could develop along the lowland corridors which lead to the main centres of population and industry.

All the gas landed need not, of course, be used as fuel. Several proposals for establishing gas-based petrochemical factories have been mooted for the Peterhead area, and more may emerge before long. At the time of writing, January 1975, an application for one specific plant is under consideration by the local planning authority. Scanitro, a consortium formed by Norsk Hydro of Norway and Supra of Sweden, announced plans in November 1974 for the setting up of a £50 million liquid ammonia plant close to Peterhead (Fig. 6.9). Scottish Agricultural Industries, a subsidiary of ICI, could become a partner in the operation. The proposed unit would export the bulk of its production by sea from a terminal in Peterhead Bay, to which it would be connected by pipeline.

The scheme has raised considerable objections locally, including representations from Peterhead Town Council that the proposed plant is wrongly located, downwind of the town and outwith areas specially zoned for industry as recently as 1973. Fears that the plant could be noisy, lead to pollution problems and even to a risk to the town from the escape of dangerous gas have also been expressed, and members of the Town Council are to visit a similar plant in Norway to judge for themselves how serious these are. It has also been suggested that establishment of one petro-chemical plant on the site could well lead, as elsewhere in Britain, to the subsequent development of neighbouring plants of a similar nature.[21]

Considering the prevailing winds the proposed site looks less than ideal. The question arises of whether it is vital for the factory to be quite so close to the shipment point. Could a longer pipeline reasonably be installed if a different site in the Peterhead area were found for the plant? There seems little doubt that petrochemical plants are likely to be sited in the general area of the gas landfalls, so approval of the site for the first one must be given special care. Wherever it is sited it must add to the already severe strain on the infrastructure of the Peterhead area, and therefore to the immediate financial problems of a local authority which is already over-burdened with the task of providing roads, drainage and housing to serve a multiplicity of major developments.

Oilfield servicing at Peterhead makes use of the former Harbour of Refuge which was constructed in Peterhead Bay by prison labour between 1886 and 1956. This sheltered 121 ha (300 acre) expanse of deep water, strategically located at the easternmost tip of mainland Scotland, attracted the attention of oil developers at an early stage. In November 1971 Peterhead Harbour Trustees gave a 99-year lease to Arunta (Scotland) Ltd of 1.8 ha (4.5 acres) of Keith Inch, a largely derelict island which sheltered the town's harbours. The company planned to use the site for a major custom-built oil service base, with a new jetty projecting into the Harbour of Refuge. They soon found, however, that the Act of 1884 by which that harbour had been established

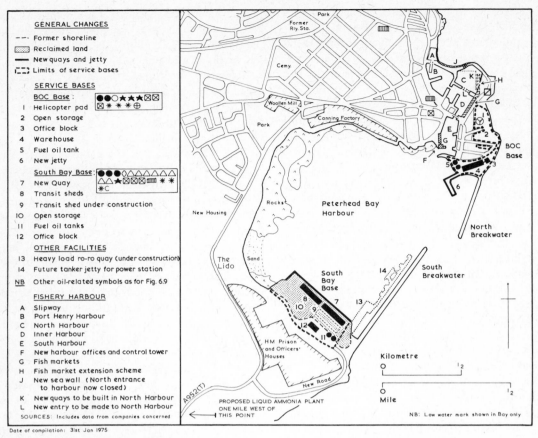

GENERAL CHANGES
- – – Former shoreline
- ▨ Reclaimed land
- ▬ New quays and jetty
- ⌐⌐ Limits of service bases

SERVICE BASES
BOC Base : ●●○★★★⊠
1 Helicopter pad ⊠✳✳✳⊕
2 Open storage
3 Office block
4 Warehouse
5 Fuel oil tank
6 New jetty

South Bay Base : ●●○△△△△△
7 New Quay △△★★⊠⊠⊞✳✳
8 Transit sheds ✳C
9 Transit shed under construction
10 Open storage
11 Fuel oil tanks
12 Office block

OTHER FACILITIES
13 Heavy load ro-ro quay (under construction)
14 Future tanker jetty for power station
NB Other oil-related symbols as for Fig. 6.9

FISHERY HARBOUR
A Slipway
B Port Henry Harbour
C North Harbour
D Inner Harbour
E South Harbour
F New harbour offices and control tower
G Fish markets
H Fish market extension scheme
J New sea wall (North entrance to harbour now closed)
K New quays to be built in North Harbour
L New entry to be made to North Harbour
SOURCES: Includes data from companies concerned

Date of compilation: 31st Jan 1975

FIG 6.9   PETERHEAD HARBOUR

---

prohibited the building of any new structures within the bay, so their scheme could not get started. In the spring of 1972 though, another firm, Site Preparations Ltd of Barrhead, proposed construction of berths on the opposite side of the bay for the servicing of large oil rigs, and there were indications that further firms had ideas for developments in the same general area.

In the face of an increasingly complicated situation, and concerned that much of the rig servicing trade might be lost by default to foreign bases, the Scottish Office acted remarkably quickly. Since 1960 responsibility for the regulation of the Harbour of Refuge had been vested in the Secretary of State, through the Department of Agriculture and Fisheries. In only a few weeks before Parliament rose for the summer recess of 1972, a new Act[22] giving the Secretary of State for Scotland powers to 'develop, maintain and manage, or allow others to do so, harbours in Scotland made or maintained by him' was passed through all its stages. By the end of August initial details were released of the Government's intention to use the powers itself

to reclaim land and construct quays at the south side of the bay. In mid-November it was announced that these facilities would be leased on a 15-year basis to Aberdeen Service Company (North Sea) Ltd, a subsidiary of Sidlaw Industries of Dundee. Permission to build their jetty at the north-east corner of the bay was given to Arunta (Scotland) near the end of October and its construction started almost immediately. Reclamation for the Government quays began in January 1973.

Since then the development of both bases has been extremely rapid. The Arunta base, since renamed Peterhead (BOC) Base,[23] was partly in operation in the spring of 1973, using temporary accommodation in the fishery harbour alongside to service lay barges working on the Forties pipeline. Fully operational since April 1974, its 4-berth jetty is now backed up by a massive warehouse, a block of offices, extensive open storage areas and a busy heliport, which provide comprehensive custom-built support for offshore engineering. At the height of the summer season in 1974 the BOC base was handling supply ships for as many as eight lay barges, and it has also become the base for the fleet of ships installing modules on BP's Forties platforms. Persistent efforts to get these into position despite repeated storms has helped to sustain the trade of the base during the winter. Withdrawal of many of the lay barges from about October to March does, however, give its traffic a strong seasonal element.[24]

Expansion of Aberdeen Service Company's South Bay Base has been even more rapid. Originally planned to be brought into operation by stages, with berths for only four ships being provided initially, it has attracted so much traffic that the developers have gone straight ahead with the entire works programme. They are currently trying to get additional land for back-up purposes about 6 km south of the harbour. Although also involved in the more seasonal traffic of servicing the installation of production facilities, a larger proportion of ASCO's customers are undertaking exploratory drilling. Occidental, Phillips Petroleum and the Burmah Oil Development Company are all using this as their main centre of marine operations.

The 457 m (1 500 ft) long concrete quay has a minimum depth of 7.3 m (24 ft) alongside and can berth seven large supply vessels at a time, each fully serviced on the 'one-stop' principle. Between January 1973 and June 1974 10 ha (25 ac) of reclaimed land was created behind the quay, largely built up by sand dredged from the sea bed a short distance to the north of Peterhead. Two large transit sheds back the quay, with a third under construction (Fig. 6.9), and substantial areas are used for open storage. In October 1973 a semi-submersible rig, *Bluewater 3*, tied up at the quay for inspection and to take on stores between drilling assignments, and larger rigs could if necessary be moored for servicing in the deeper central areas of the bay.

Large pipelaying and bury barges are in fact regular visitors to the harbour, though not to either of the bases. During the stormy autumn and winter of 1974 Peterhead Bay was frequently crowded with such vessels and their attendant supply ships. Additionally, other lay barges and the huge heavy-lift ships and module-carrying barges assembling the platforms for the Forties Field have lain off the coast between Peterhead and Aberdeen for days at a time when bad weather has interrupted their operations.

Creation of the quay and reclaimed land for Peterhead South Bay Base was undertaken by the Government at a cost of around £2½ million. The operating company is spending a further £2 million on the provision of the necessary buildings, roads, storage units and equipment. Works yet to be completed

as part of the Government scheme for the harbour are a general purpose quay at which units of machinery up to 400 tons in weight will be delivered by ro-ro vessels for installation in the new Peterhead power station, and a jetty to berth 40 000 ton tankers. From this jetty fuel oil will be transferred to the power station when it comes into operation, about 1977.[25]

While overall control of developments in the former Harbour of Refuge is vested in the Secretary of State for Scotland, he has set up a body representative of a number of local interests to take responsibility for running the new port. This body, Peterhead Bay (Management) Company, exists alongside the Peterhead Harbour Trustees who run the fishery harbours. It co-operates with them and fishery interests through a special liaison committee and employs the Trustees' harbourmaster to supervise navigational control of traffic within the bay. The very big upsurge in white fish landings at Peterhead in recent years has increased the traffic of fishing boats through the north end of the bay.[26] As at Aberdeen a radar-assisted system of navigational control has had to be installed, in this case in new harbour offices specially built to overlook both harbours. (Fig. 6.9).

C: Montrose

At Montrose, as in Peterhead, land reclamation using dredged materials has created new port facilities specially designed to cater for rig service vessels. In this case the initiative was taken by Montrose Harbour Trust with the support of a company of service base developers, Sea Oil Services, to whom the land and quays have been leased.[27] The construction of these facilities was pushed ahead very rapidly, but unlike ports further north oil-related traffic has been slow to develop.

Fig. 6.10 indicates how 15½ ha (38 ac) of new land was added to Rossie Island, extending it eastward and eliminating the former south channel of the River South Esk. The 305 m (1 000 ft) long quay and extensive open storage on the base is backed up by a further area in a former railway yard just north of the harbour. Although the first boat was able to use the South Quay in May 1974, and the full berthage has been available since the late autumn, the base is at present distinctly under-utilised. In January 1975 only eight service boat movements were recorded at the harbour, making a total of 82 in the financial year to date.

Two rigs are now being serviced from the base and plans have been passed for extending both the warehouse space and Sea Oil's engineering workshop.

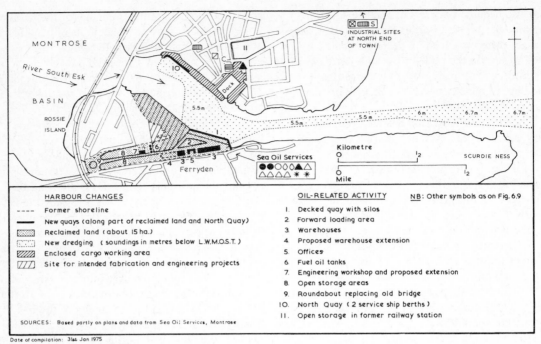

FIG 6.10   MONTROSE HARBOUR

Considerable emphasis is put by the company on the use of this workshop to manufacture oilfield tools and equipment under licence from North American firms. There has been no development yet, however, of a projected module construction yard on the northern section of the reclaimed land, where an extension of the dredged channel was provided to allow barges to come alongside for loading structures.

While the financial results to date must be disappointing to the developers the greatly improved shipping channel has given a big impetus to the commercial trade of Montrose Harbour,[28] and a number of small oil-related businesses have been set up in the town. In total the base and these firms employed about 180 people at the end of 1974, roughly double the number involved six months previously.[29] With three oil companies and several drilling and specialist engineering companies now established in and near the base the potential for growth is certainly present at Montrose, but the return on investment is taking much longer to materialise than at ports closer to the concentration of offshore activity off North East Scotland and the Northern Isles.

D: Dundee and District

Oil-related activity at Dundee, although much greater

than that at Montrose, has also grown relatively slowly. There are two oil company bases at the harbour, both gradually extending their operations and attracting around them an increasing number of supply and servicing firms,[30] but substantial further development in this field would seem to depend upon oil exploration companies coming back to explore more fully the potential of blocks in the central basin of the North Sea. For the present there are clear signs that, to increase effectively its share of the new business available, the area needs to attract further oil engineering concerns. In the latter half of 1974 numbers directly employed in oil-related work at Dundee increased from about 340 to 430, but at Arbroath the work force of the recently established Halliburton Manufacturing and Services factory rose from about 20 to 230.[31] Unlike Aberdeen, Dundee is basically an industrial centre with a distinctly smaller, though growing, service sector. Contraction of the jute industry in recent years has meant a substantial decline in imports of raw jute and has left quay space at the harbour increasingly under-utilised. However, despite the crowded situation at Aberdeen, the oil servicing industry has on the whole shown itself reluctant to expand greatly at Dundee.

Possible reasons for the limited involvement of the oil industry in the city include the lack of a proper airport and the greater sea distance from oil exploration areas than that of ports in the North East (Fig. 5.2), also the relative delay in developing a co-ordinated publicity machine to present the city's attractions to developers. Of these, the most significant factor is undoubtedly the lack of an airport comparable to Dyce. Dundee's airstrip at Riverside has a feeder service to Glasgow, and is regularly used by charter aircraft, but it lacks the capacity to handle the types of plans generally in use on scheduled national and inter-European services and no agreement has yet been reached locally (January 1975) about proposals for extending and upgrading it. Oil companies considering where to establish large concentrations of their executive staffs demand ready access to air services with direct flights to London and other main centres of the industry. By now also there is such a wide range of administrative and supply concerns established in Aberdeen that firms may derive significant economies by locating there alongside them.[32]

Fig. 6.12, drawn on the same scale as Fig. 6.8, shows both the much more restricted level of development at Dundee and how closely this is confined to the harbour area. First in the field was BP's marine base which was established at Queen Elizabeth Wharf in 1971 (Fig. 6.13). It can service three boats at a time and is backed up by storage areas at the east end of the docks and in part of the station yard to the west, as well as by a strip of land reclaimed from Camperdown Dock just behind. When drilling work starts on the production platforms of the Forties Field, BP will operate supply services to them from Dundee, as well as to the three exploration rigs currently drilling for the company.

Further east, beyond the main commercial wharf of the port, the former boiler shop and a fitting out quay of the Caledon shipyard have become the base for Ocean-Inchcape Ltd, a firm which operates service vessels from a variety of ports around the North Sea, and has recently commissioned a geophysical survey ship. Run by Ocean-Inchcape (Caledon) Ltd—formerly called Tayside Offshore Supplies and Services—the base has attracted a number of tenants, but none so far in the exploration, drilling or offshore engineering categories, which would be likely to lead to substantial development. With berth space for only one ship at a time the scope is in any case limited.

In contrast, the Dundee Petrosea base immediately alongside has quay space for three service ships and will have a fourth berth when Dundee Harbour Trust complete the second stage of their Princess Alexandra Wharf. Like Ocean-Inchcape (Caledon), Dundee Petrosea links a local firm with the expertise of an established base and service ship operator, in this case

the Singapore Company Thiess Petrosea International. The base is leased in its entirety to the oil exploration firm Conoco and to firms working in association with it.

Dundee Harbour Trust's main investment in the oil industry to date has been the reconstruction of the western half of their Eastern Wharf. 180 m (600 ft) of new quay were completed in 1973 and renamed Princess Alexandra Wharf. The remaining 76 m (250 ft) should be finished by the end of March 1975. Depths of 6 m (20 ft) LWS are available here and 7.6 m (25 ft) at Queen Elizabeth Wharf. Preliminary work has also been undertaken for a much more ambitious provision of deep water quays and storage space at the Stannergate, just east of the Caledon Shipyard. This scheme could involve about 457 m (1 500 ft) of wharfage with a depth alongside of 12 m (40 ft) at low tides, and by reclamation would provide about 10 ha (25 ac) for transit sheds, open storage and so on.

Surveys have found no physical limitations to obtaining this depth either off the Stannergate or in the approaches to the harbour, which at present in places have limits of 5.5 m (18 ft) at the lowest tides. However, before embarking on such an expensive project the Harbour Trust must first ensure that one or more oil companies would be prepared to make use of the facilities created. Recent forecasts have suggested a limited demand for new bases in the next few years, except perhaps in Orkney or Shetland,[33] and the scheme would require Government approval through the National Ports Council, so there must be some doubt as to whether it will be undertaken.

If, as seems likely, Dundee's main scope for participating in the oil industry lies in the further development of specialist oilfield engineering, cargo facilities at the harbour could prove an attraction for incoming firms. Merchant ships of up to 20 000 tons deadweight can be handled at King George V Wharf, and the port has already had some direct shipments of oilfield equipment from US Gulf ports. Other locational advantages to be stressed in promoting industrial development include the well-established engineering tradition in Dundee and district, its plentiful and relatively cheap building land, good road and rail links, and the availability of labour following reorganisation of local textile and engineering firms.

Tayside Development Authority are in fact giving priority to reviewing the likely demand for labour and the types of training which need to be provided. Meanwhile, the exceptionally quick growth of the Halliburton engineering operation at Arbroath has been made possible by the ready availability of labour.

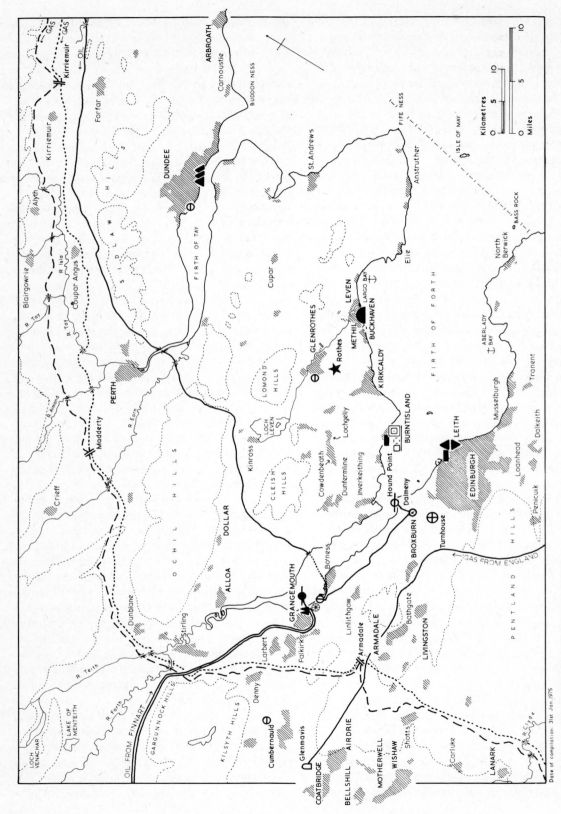

FIG 6.11 TAYSIDE AND FIRTH OF FORTH

Date of compilation 31st Jan 1975

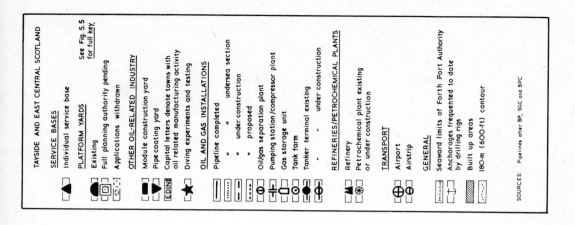

In addition to their present factory the company has bought another modern unit on the industrial estate at the west end of Arbroath, along with 11 ha (27 ac) of land for the erection of further buildings. The company has moved its accounting and computer groups to this estate from London, and its forecast of having a total of 600—800 employees at Arbroath begins to look fully credible.

Research and development work which points the way to an increased future involvement of the area in the oil industry includes work done by the Department of Electrical Engineering at Dundee University to provide satellite cloud cover pictures of the North Sea for detailed weather forecasting, and the staffing by local doctors and nurses of a special pressurised hospital unit at Dundee Harbour for the treatment of divers. Additionally, a precision engineering firm in Perth has developed its own designs of equipment for mini-submarines and hopes to establish a regular business in servicing the growing number which are operating in the North Sea.

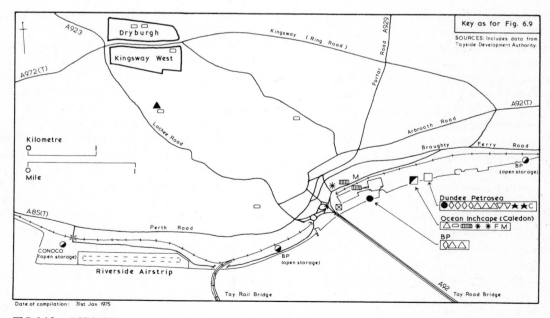

FIG 6.12   DUNDEE

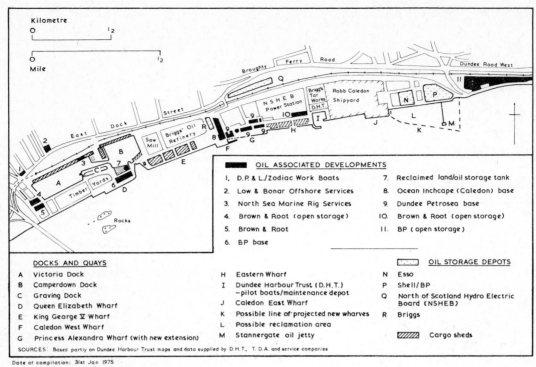

FIG 6.13   DUNDEE DOCKS

The following is the content within the figure:

OIL ASSOCIATED DEVELOPMENTS

1. D.P. & L./Zodiac Work Boats
2. Low & Bonar Offshore Services
3. North Sea Marine Rig Services
4. Brown & Root (open storage)
5. Brown & Root
6. BP base
7. Reclaimed land/oil storage tank
8. Ocean Inchcape (Caledon) base
9. Dundee Petrosea base
10. Brown & Root (open storage)
11. BP (open storage)

DOCKS AND QUAYS

A Victoria Dock
B Camperdown Dock
C Graving Dock
D Queen Elizabeth Wharf
E King George V Wharf
F Caledon West Wharf
G Princess Alexandra Wharf (with new extension)

H Eastern Wharf
I Dundee Harbour Trust (D.H.T.) –pilot boats/maintenance depot
J Caledon East Wharf
K Possible line of projected new wharves
L Possible reclamation area
M Stannergate oil jetty

OIL STORAGE DEPOTS

N Esso
P Shell/BP
Q North of Scotland Hydro Electric Board (NSHEB)
R Briggs

Cargo sheds

SOURCES: Based partly on Dundee Harbour Trust maps and data supplied by D.H.T., T.D.A. and service companies

Date of compilation: 31st Jan 1975

6. Firth of Forth

With its ports lying still further south and west than those of the North East and Tayside, the Firth of Forth has been less immediately attractive as a servicing base for rigs operating in the northern North Sea. It has, however, proved to have a number of other attractions for the oil industry which are not so easily matched by areas further north. The spaciousness of its estuary has, for instance, made it a haven for semi-submersible rigs. Accessibility to established areas of heavy industry and to industrial labour has encouraged the growth of offshore engineering. Dry dock facilities have attracted rig service vessels for repairs and refitting, and substantial spare capacity at deep water berths in its ports has attracted both an increased commercial traffic and the establishment at Leith of a pipe coating plant. Perhaps most critical of all, the Forth, as the site of the only existing oil refining and petrochemical complex in Scotland, is to become the first centre north of the border to refine and export North Sea crude oil.

No one would pretend that anchorages in the Firth of Forth offer ideal shelter for the repair and refitting of large oil rigs, but there have been few alternatives available on this side of the North Sea (p. 78), and the bulk of refitting of semi-submersibles working in the North Sea has gone by default to continental ports. Largo and Aberlady bays have, however, several times proved to be vital refuges for rigs which have been damaged in storms.[34] A considerable number of rigs have also anchored off Fife between Elie and Buckhaven to offload or take on stores between drilling assignments, and more recently a very large rig, newly-built in Germany, spent a month there for the final stages of the building of its superstructure. Such work has drawn on engineering labour from Central Scotland as well as providing contract work for service vessels operating out of Dundee and Forth ports. Local residents have objected to the noise of operations being carried out on the rigs but there is likely to be a continuing demand on these anchorages. Following objections from fishermen who use the area it has been agreed that Aberlady Bay will be used for rigs only when anchorages on the north side of the firth are full, so there could be an increasing pressure on the Fife coast.[35]

Availability of suitable labour has also been a factor in attracting offshore engineering work to sites

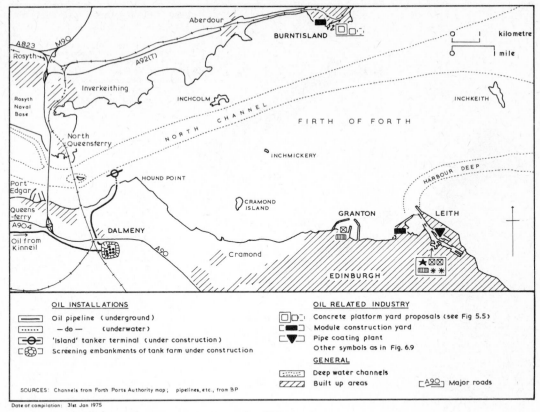

Legend within figure:

OIL INSTALLATIONS

Oil pipeline (underground)
— do — (underwater)
'Island' tanker terminal (under construction)
Screening embankments of tank farm under construction

OIL RELATED INDUSTRY

Concrete platform yard proposals (see Fig 5.5)
Module construction yard
Pipe coating plant
Other symbols as in Fig. 6.9

GENERAL

Deep water channels
Built up areas
A90 Major roads

SOURCES: Channels from Forth Ports Authority map; pipelines, etc., from B P

Date of compilation: 31st Jan 1975

FIG 6.14   HOUND POINT, BURNTISLAND AND LEITH

on land round the Firth of Forth. In October 1972 Redpath Dorman Long (North Sea), a consortium formed by the British Steel Corporation and three Italian underwater engineering firms, took over the former Wellesley Colliery alongside Methil Docks. By transforming the site into a 49 ha (120 ac) yard for steel platform production, the company both rehabilitated a derelict area and provided much needed jobs in a district of considerable manufacturing and mining unemployment. Its first order, a 3 500 tonne jacket for the Auk Field, was floated out on a barge in July 1974 and a 12 000 unit for Shell's Brent Field is currently almost complete. Up to 800 men have been employed by the yard at peak periods.

Similarly, at Burntisland (Fig. 6.14) Robb-Caledon have filled in a tidal basin at their former shipyard, and have transformed it into a thriving module construction yard employing over 200 men. Most of the work is carried out under cover in a huge fabrication shed which was erected around the module assembly work while it was in progress. Buildings and equipment of the former shipyard are also used, and barges can be brought alongside the quay where big cranes, previously used in fitting out ships, can assist in loading the modules

On the opposite side of the firth the Motherwell Bridge (Offshore) Ltd module construction yard at Leith occupies an open site on land reclaimed from the Western Harbour. By virtue of its position on the south side of the estuary this yard can also serve as an assembly and shipment point for heavy equipment produced by the group's heavy engineering plants at Motherwell and Broxburn, while the big expanse of sheltered water in the Western Harbour provides a safe anchorage for loaded barges waiting for suitable weather conditions for the installation of modules on platforms.

Leith Docks (Fig. 6.15) have in fact proved attractive in a number of ways to firms engaged in offshore engineering. A major reconstruction scheme completed in 1969 made the entire dock complex an enclosed one, with lock gates operating on a 24-hour basis and capable of handling ships of up to 40 000 tons deadweight. Considerable areas within the harbour are

99

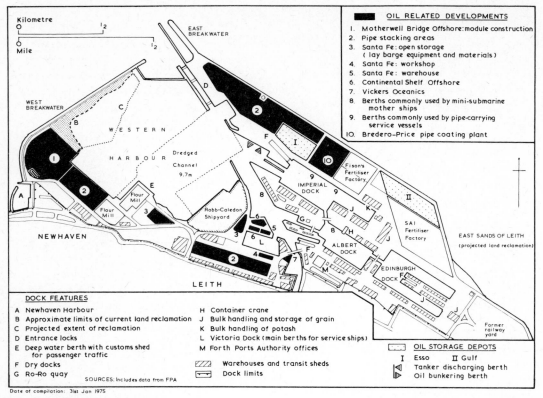

Kilometre
0 | | | 2
0 | | | 2
Mile

EAST BREAKWATER

WEST BREAKWATER

C

B

WESTERN HARBOUR

Dredged Channel 9.7m

A

1

2

2

E

Flour Mill

Flour Mill

3

Robb-Caledon Shipyard

NEWHAVEN

D

2

F I

10

Fison's Fertiliser Factory

9

IMPERIAL DOCK

9

II

SAI Fertiliser Factory

EAST SANDS OF LEITH
(projected land reclamation)

8

J K

3

6

5

G

B H

J

3 6

L

F

C

ALBERT DOCK

4

7

M

EDINBURGH DOCK

F

J

J

A

Former railway yard

2

LEITH

OIL RELATED DEVELOPMENTS
1. Motherwell Bridge Offshore: module construction
2. Pipe stacking areas
3. Santa Fe: open storage (lay barge equipment and materials)
4. Santa Fe: workshop
5. Santa Fe: warehouse
6. Continental Shelf Offshore
7. Vickers Oceanics
8. Berths commonly used by mini-submarine mother ships
9. Berths commonly used by pipe-carrying service vessels
10. Bredero-Price pipe coating plant

DOCK FEATURES
A Newhaven Harbour
B Approximate limits of current land reclamation
C Projected extent of reclamation
D Entrance locks
E Deep water berth with customs shed for passenger traffic
F Dry docks
G Ro-Ro quay

H Container crane
J Bulk handling and storage of grain
K Bulk handling of potash
L Victoria Dock (main berths for service ships)
M Forth Ports Authority offices

Warehouses and transit sheds
Dock limits

SOURCES: Includes data from FPA

OIL STORAGE DEPOTS
I Esso   II Gulf
Tanker discharging berth
Oil bunkering berth

Date of compilation: 31st Jan 1975

FIG 6.15 LEITH DOCKS

given over to industries such as grain milling and fertiliser manufacture which process bulk cargoes unloaded immediately alongside. Progressive reclamation of land from the adjacent Firth has allowed a spacious layout of roads and storage areas and a feasibility study into further reclamation of about 162 ha (400 ac) from tidal flats to the east of the docks is currently in progress.

This capacity to provide extensive storage and processing space, and the ability of the port to handle big freighters bringing in steel tubes, iron ore and other necessary materials were key considerations which brought Bredero-Price (UK) to establish a pipe coating plant here in August 1972. Contracts obtained by the plant to coat pipe for the Ekofisk to Teesside pipeline and then of part of the Piper contract in turn brought pipelaying contractors to the port and established it as a base for service ships supplying the pipelaying barges. Currently (January 1975) the pipe coating plant is closing down for lack of orders, although it is hoped that this will be only a temporary set-back. However, the engineering contractors continue to be active.

Meanwhile, with ports in the North East so

heavily committed to the oil industry, Leith must remain attractive to new firms coming into the offshore business. A recent indication of this was the establishment by Vickers Oceanics of a base in the docks for their fleet of three mini-submarine mother ships. This growing business of operating mini-submarines is just one example of the developing field of advanced technology subsea engineering which will be requiring land bases with good access to specialised repair and maintenance facilities. Leith's relative proximity to industrial areas in both Scotland and northern England, and its ready access to Turnhouse Airport for quick transporting of personnel and spares, should help it to attract this type of development.

This is, of course, speculation and the Firth of Forth, like other areas around the Scottish coast, has learned that not all promised or even all likely schemes come to fruition. Fig. 6.11 shows, for instance, that there have been four proposals for platform construction approved in outline by Fife County Council which have not so far materialised. An application in April 1973 by Fred Olsen Ltd for a 121 ha (300 ac) site at Buckhaven was abandoned after fuller examination of the site revealed geological difficulties, and plans lodged in June 1973 by ANDOC, an Anglo-Dutch consortium, for a development at Burntisland were withdrawn in October 1974 in favour of a site at Hunterston on the Clyde (p. 105). The same site, a former railway yard of 6.9 ha (17 ac) alongside Burntisland's East Dock, was again applied for in 1974; this time by two separate concerns which had each prepared special plans for building concrete platforms in the 27m (90 ft) depth of water available between the shore and the main shipping channel. Berg Platforms claimed to have a special light-weight aggregate which could revolutionise concrete platform construction, while Caledonian Platform Structures claimed to have produced a design which could be completed to its full height only 750 m offshore. Fife County Council gave outline planning permission to both projects, and in November 1974 the Forth Ports Authority, faced with choosing between them, gave a one-year option on the harbour site to Caledonian Platform Structures.[36] Both schemes originated in Scotland and were put forward by Scottish registered companies. Prospects of a possible 500 new jobs at Burntisland now rest with Caledonian Platform Structures obtaining a contract for their design, a necessary prerequisite to getting approval for the project from the Scottish Secretary.

Meanwhile, in and around Edinburgh itself a substantial amount of new business has been won by marine, electrical and electronics engineering firms, which have turned their attention to the new offshore market, as well as by a branch at Livingstone of a US firm which has a specialist interest in oilfield technology. Provision of full Development Area status for Edinburgh and Leith in August 1974 has put the area on a much fairer footing in competing for business with firms elsewhere in Scotland, and it has been further encouraged by government decisions in January 1975 to set up the first official school of drilling technology at Livingstone and to finance a petroleum engineering centre alongside the Heriot-Watt University Institute of Offshore Engineering on the western outskirts of Edinburgh (pp. 62–63).

Depths of 24 m (80 ft) and more, which obtain from about the Forth Bridges eastward, have been vital in the locating of the new Hound Point tanker terminal which lies at the end of the oil pipeline from BP's Forties Field (Fig. 5.6). The necessary combination of deep water close inshore, adequate for 250 000 tonne tankers and with sufficient shelter for their safe berthing, occurs only for short distances upstream and downstream of the Bridges, and it is obvious that large tankers should avoid having to pass beneath them if at all possible. Such factors must have weighed heavily with the Secretary of State in permitting the construction of the terminal and its associated tank farm at Dalmeny within Edinburgh's green belt. Exceptional care has been taken to reduce to a minimum the environmental impact of both units. The terminal has been built as an artificial island on piles, with the oil loading and ballast discharging pipes buried under the bed of the estuary,[37] and the high embankments which screen the tank farm were created in part from an old oil shale bing which covered part of the site.

The need for the terminal in the first place arises from the fact that BP's refinery at Grangemouth must balance its intake of light Forties crude with an equal amount of heavier crudes brought in through Finnart; and also from the apparent reluctance of the company to expand its oil refining capacity in Scotland beyond the needs of the Scottish domestic market (p. 46). With the refinery taking no more than $4\frac{1}{2}$ million tonnes per year of the Forties' output, up to 15 million tonnes per year of crude oil will be exported through Hound Point, to refineries elsewhere in Britain or abroad. The terminal should be complete and ready for use by the middle of 1975.

At Grangemouth a special plant has had to be developed which will separate gases from the crude oil to make it safer for transporting by tanker. At one stage negotiations were in progress with the South of Scotland Electricity Board for the use of the recovered gas as fuel in Kincardine power station on the opposite side of the Forth, but the idea of converting this coal-burning unit to dual firing was opposed by the British Gas Corporation[38] which has a prior right to purchase all natural gas which is to be used as fuel (p. 30). It must in any case be questioned whether government fuel policy in the changed circumstances of the 1970s could allow replacement of coal by natural gas in a nationalised power station.

BP's policy is now to liquefy the heavier gases, propane and butane, in the existing refinery and to export them from a terminal in Grangemouth Docks. The bulk of the lighter gases will be used as fuel within the refining and petrochemical plants at Grangemouth.[39] Such a policy of burning the gases may appear wasteful but it is normal practice to fuel refinery processes in this way. Rezoning for industry in September 1974 of 243 ha (600 ac) of farmland immediately to the west of Grangemouth Docks raises the question of the possible establishment there of industrial plant which could provide a market for petroleum gases to be used as raw material. While sale of gas by BP for such a purpose would require government approval the British Gas Corporation would not have any right to object.

7. West Central Belt and the Firth of Clyde

Although over 40% of the total population of Scotland live in the West Central Belt[40] and despite its being indisputably the industrial heart of the country, this area has so far had a less than proportionate share of the developments associated with offshore oil and gas (Table 5.6). This is a matter of concern for economists, and still more so for politicians because the area has for long suffered chronic unemployment[41] despite persistent efforts to establish new growth industries alongside its declining traditional ones. This situation is, however, readily understood if one considers the typical economic geography of the earlier stages of developing an off-shore oil province. Further, the picture today is not by any means so completely negative as might appear at first sight.

Geophysical and test drilling programmes offshore necessarily put pressure on the coastal areas immediately accessible, in this case the east coast. Discovery of oil or gas brings the need for production platform and pipeline construction, but for reasons given earlier (p. 56) firms building these also had good reason to favour the east of Scotland for their projects. Even for concrete platforms, interest in sites on the Clyde came relatively late, largely when it became clear that deeper water sites further north were not to be made available (pp. 106 & 116).

In other words, the offshore oil and gas industry was not geographically constrained to come to West Central Scotland and the estuary of the Clyde until quite a late stage in its development. Further, the trade which it has brought to the industries of Glasgow and the surrounding area has been largely for the normal line of products which the region has been accustomed to produce. For this reason oil-related developments in the area have not been immediately obvious to the public at large. Because they have not brought striking changes to the area they have failed to attract the interest or the publicity which has attended developments on the east coast. Their significance so far has lain largely in the extent to which they have helped to limit unemployment and further plant closures rather than in their creation of new firms and new industries. It is important, however, to point out that many recent oil-related industrial developments in the area are not only at the forefront of technology but are also on such a scale that if they had occurred elsewhere they would have been singled out as of remarkable economic and social significance.

West Central Scotland is, in fact, particularly well-fitted to supply offshore engineering requirements (a characteristic which it shares with North East England). Its steel industry has been traditionally orientated to production of heavy plates and sections such as are needed by shipbuilders and construction engineers. It is also a special producer of steel tubes up to 46 cm (18 cm) in diameter. The heavy engineering works of the area have most of the machinery and skills needed in offshore fabrication work. Many of them make products such as gas turbines, cranes, valves and pumps which are directly needed by the oil industry, and a very similar situation obtains among shipbuilders and their associated suppliers. In short, the basic capacity to supply many of the needs of the oil industry exists in West Central Scotland, and the questions for consideration are how actively firms have sought out potential business and the extent to which they have been prepared to undertake research and development work to tailor their products to the special needs of the industry.[42]

Particular credit attaches in this respect to the British Steel Corporation. In the past six years BSC has produced 90% of the well-casing and steel tube needed in the drilling programmes of the UK sector, much of it from the Clydesdale works at Bellshill (Fig. 6.11). Reconstruction of this works, now nearing completion, has doubled its capacity replacing outdated open-hearth furnaces with modern electric ones which give better quality steel. Special quality control systems for casing pipes have been introduced, and both the Clydesdale and a related Coatbridge works have been equipped to turn out tubes with a new French-designed casing coupling which is competitive with standard US designs. Further expansion of all this capacity is under consideration to meet the sharply rising demand as multiple production wells begin to be drilled in the North Sea.[43]

In similar fashion Hallside steelworks near Cambuslang now produces a specially researched 'node' steel with welding properties designed to overcome difficulties previously experienced in producing key nodes (or leg junctions) for steel platform jackets. Much of the fabrication of these nodes is undertaken at BSC steelworks in Glasgow and at Glengarnock in Ayrshire (Fig. 6.16). Rising demand for high-quality heat-treated plates for oil storage tanks, steel piles, cranes and so on is being met by nearly quadrupling the tempering furnace capacity at Clydebridge Steelworks, Cambuslang; and the high-strength plate mill capacity at Dalzell works in Motherwell has been increased by over 1 000 tons per week.[44] Additionally, a plant involving a new refining process for producing heavy plate to atomic vessel standards is under construction at Dalzell which will

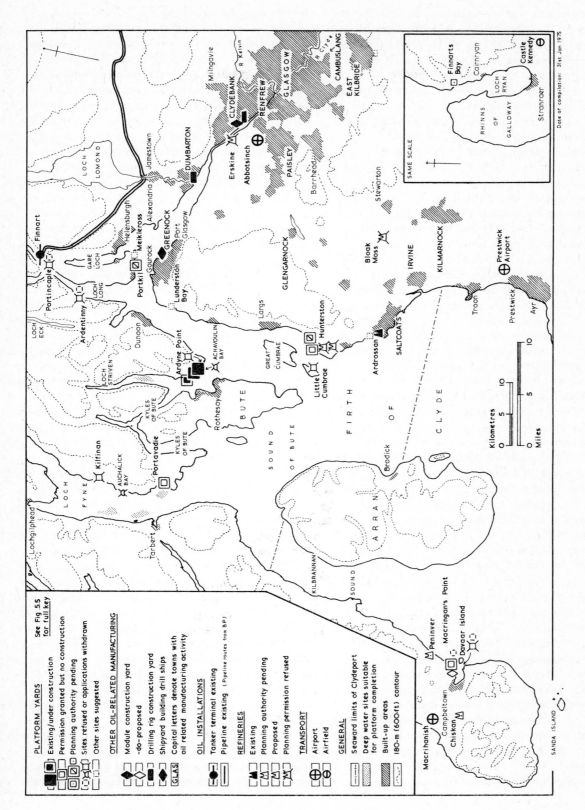

FIG 6.16   FIRTH OF CLYDE

PLATFORM YARDS   See Fig 5.5
                        for full key
■ Existing/under construction
□ Permission granted but no construction
◨ Planning authority pending
◫ Sites refused or applications withdrawn
⊡ Other sites suggested

OTHER OIL-RELATED MANUFACTURING
◆ Module construction yard
◇ -do- proposed
▬ Drilling rig construction yard
◆ Shipyard building drill ships
GLAS Capital letters denote towns with
      oil related manufacturing activity

OIL INSTALLATIONS
● Tanker terminal existing
━ Pipeline existing (Pipeline routes from BP)

REFINERIES
▲ Existing
▲ Planning authority pending
▲ Proposed
⋈ Planning permission refused

TRANSPORT
⊕ Airport
⊖ Airfield

GENERAL
─·─ Seaward limits of Clydeport
⠂⠂ Deep water sites suitable
     for platform completion
▨ Built-up areas
⋯ 180·m (600·ft) contour

SANDA ISLAND

Date of compilation: 31st Jan 1975

SAME SCALE

Kilometres 0   5   10
Miles   0   5   10

eventually be able to turn out slabs up to 50 tonnes in weight. Steel shortages have plagued offshore work and fabricating in the past, but the British Steel Corporation in Scotland is investing heavily to try to meet the requirements.

Firms engaged in heavy and medium engineering which have undertaken substantial amounts of oil-related business range right across the West Central Belt from Coatbridge and Motherwell in the east to Clydebank, Renfrew and Irvine in the west. It has not proved possible in the present study to document fully their location or the extent of their activities,[45] but towns with firms of this type are identified on Figs. 6.11 and 6.16. An appreciable number of firms report oil-related business to an annual value of £1 million or more. Some, such as Martin Black, wire rope manufacturers of Coatbridge, have invested sums of this order in new capital equipment specially directed to produce goods for the oil market. In a number of cases experience previously gained in work done on nuclear engineering contracts has stood firms in good stead, and a few already had specific experience in oil or petrochemical engineering, but the majority have been finding new outlets for their existing products and for their general expertise in mechanical and electrical engineering. So far only a small number of entirely new oil engineering businesses have been set up, but a number of companies have negotiated agreements to produce foreign-designed oilfield products under licence.

Quite a lot of the equipment produced for the oil industry will be incorporated in modules to be mounted on production platforms. Actual assembly of such modules in many ways resembles shipbuilding so it appears a 'natural' field of enterprise for Clydeside. The Foster Wheeler-John Brown Boilers module yard at Dumbarton which occupies the site of the former Denny shipyard at the mouth of the River Leven has been in operation since 1973, while vacant land alongside Clydebank's Rothesay Dock is currently being reconstructed as a module yard by John Brown Engineering (Offshore) who plan to employ 600 workers there and in another fabrication unit in the town. Substantial further suitable sites for module construction on former shipyards and disused docks exist on the Clyde from near the centre of Glasgow downstream. The fact that modules, unlike the platforms themselves, are realtively easily transportable and could therefore be exported to oilfields in other parts of the world could indicate a potential field for industrial expansion in the area. There is, however, already a shortage of skilled metal workers on the Clyde so the opportunities here may be more apparent than real.

It may well be that a substantial further provision of training schemes to produce such workers is in any case a major priority for the West Central region. Shipbuilding and engineering employers who have already lost boilermakers, welders and fitters to steel platform yards and construction contracts on the east coast are looking anxiously at the projected proliferation of concrete platform yards on the Lower Clyde.[46] Contrary to general understanding, these platforms incorporate large quantities of steel, so experienced steel workers are in considerable demand in concrete yards. As it is, shipyards on the Clyde, like the steelworks further inland, find themselves unable to recruit and retain workers in the numbers they need, despite the substantial unemployment rates in their areas.

Given this situation, and already well-filled order books for their established types of vessel, Clyde shipbuilders with one exception have been decidedly cautious so far in their entry to the offshore market. The exception, the Marathon Shipbuilding Company, took over the former John Brown, Clydebank yard from Upper Clyde Shipbuilders in 1972 specifically to build jack-up drilling rigs. With around 1 800 workers it is one of the largest single employers of oil-related labour in Scotland,[47] sharing a substantial order book with other Marathon yards in the US and Singapore. More recently, in 1974, the Scott-Lithgow group at Greenock has won orders valued at over £75 million for three advanced design drill ships, and Yarrows of Scotstoun have entered a consortium to maintain and repair rigs and platforms at sea, using a floating workshop which will be based in the Cromarty Firth.

1974 has also seen a quite remarkable outburst of interest by prospective platform constructors in sites on the Firth of Clyde. No fewer than 20 applications were submitted in the course of the year to the five counties which surround the Firth, 13 of these being to Argyll. A number of sites were applied for at least twice by different companies. At the time of writing five sites have been given final approval, three have still to be finally decided, seven have been refused planning permission by the counties concerned, and five schemes have been withdrawn by their proposers (Fig. 6.16).

The sheltered deep waters of the Firth and of the sea lochs which open into it have been the chief attraction for prospective platform builders; also the proximity of many of the sites to the potential labour force and supplies of steel, aggregate and other materials which could be drawn from West Central Scotland. The main rush of applications followed the Drumbuie Inquiry (November 1973–March 1974) when it became apparent both that applications for sites on the North West Coast would stand limited chances of success, and that the Firth of Clyde might well be deep enough for construction of a range of designs of concrete platforms. It has also been encouraged by the stated interest of the government in attracting a larger proportion of oil related work to West Central Scotland.

The key site to date, Ardyne Point, was actually chosen before these events. Application for it was first made in November 1972 and got full approval by March 1973, but its development began only in 1974. Dock construction started when the first order for a platform was received at the beginning of January 1974. At the end of that month the company got a second order and soon had permission to build a second dry dock. In May it received a third order and three months later acquired permission to build a third dock. One reason for allowing the extensions was to limit the proliferation of platform yards on the Firth, but at that stage it was made clear that no further extensions of the Ardyne site would be allowed. Because roads to it are poor and involve either long detours or ferries, materials for Ardyne Point are brought in mainly by sea, to specially constructed quays. A substantial part of the 1 000-strong labour force is also ferried across the Firth from Renfrewshire. Following initial construction in the docks of the massive bases, the structures will be moored at an adjoining jetty to have the legs formed on top and then towed either to Loch Fyne or to the Inner Sound off Kyle of Lochalsh to undergo submersion tests and have the deck units floated on top of the legs. (Until a more thorough hydrological survey is completed of the North Channel between Kintyre and Northern Ireland[48] there is some doubt as to whether fully completed platforms can be navigated out of the Firth of Clyde.)

No other platform yards are yet in production on the Clyde, but on 10th January 1975 the Scottish Secretary gave permission for three others to proceed. The first, Portavadie on Loch Fyne, had been the subject of a public inquiry in September 1974, and was approved despite a separate consultants' report to the government which recommended that Loch Fyne should be declared a conservation area with no fixed sites permitted round its shores.[49] Water immediately alongside the proposed yard is about 30 m (100 ft) deep and other points in the loch are as much as 213 m (700 ft) so all stages of construction can be undertaken close together here, provided the structures can then negotiate the 46 m (150 ft) bar at the mouth of the Firth of Clyde

and the passage through the North Channel.

The other two sites approved on 10th January 1975 were Macringan's Point near Campbeltown and a yard at Hunterston in Ayrshire. At Campbeltown Mowlem/Taywood Seltrust, the unsuccessful candidates for the Drumbuie scheme, were given permission to build a modified version of their Condeep platforms, along with concrete semi-submersible drilling rigs. At Hunterston an application by the Anglo-Dutch Offshore Concrete Group (ANDOC) received approval. Another consortium which had a longer standing interest in the site, but unlike ANDOC did not have a preferred platform design, was refused permission to go ahead unless it could land an order for its proposed type of platform.[50] Work on preparing the Portavadie yard is to go ahead almost immediately but at the end of January 1975 there was no indication of similar action at Campbeltown or Hunterston. After a year of feverish planning activity for platform yards on the Firth of Clyde the question currently has become one of whether there will be sufficient orders to justify the establishment of further active sites. This question must also hang over the proposals for a yard at Portkil, opposite Greenock. These were examined by public inquiry in January 1975 so it will be some time before the Secretary of State can announce a decision. As things stand (early February 1975), six out of seven groups which have government approved concrete platform designs have been given full planning permission for yards in Scotland. One of these, Howard-Doris, which is in process of developing the Loch Kishorn site in Wester Ross (pp. 106–108), has lodged an application for a second yard at Finnarts Bay near Stranraer. It is understood that the firm would use this site to double their capacity for building the initial stages of concrete platforms. The structures would be towed to Loch Kishorn for completion.

8. The North West

The fjord coastline of the west coast with its deep water close inshore is attractive to the oil industry on several counts. Of immediate interest are facilities for the testing of underwater equipment and the training of divers in deep water, also the location of some of the best sites in Europe for offshore engineering, particularly the construction of production platforms of concrete gravity design. Further, should petroleum be discovered in quantity in sea areas to the west of the mainland or the Hebrides, there is every likelihood that most of the servicing of rigs and platforms will be done from West Coast ports (Fig. 5.2).

Since March 1972, the waters and sea bed of Loch Linnhe have been used by Underwater Engineering Group Trials Ltd (UEG) for testing a wide range of underwater equipment. The work is being undertaken in water depths ranging from 80–225 m (260–740 ft), that is, comparable in depth to those areas of the North Sea where drilling and preparations for the production of petroleum are currently under way. The contrast in the physical conditions between the inner and outer parts of the loch make Loch Linnhe particularly well-suited for the conduct of experiments and the testing of equipment under a wide variety of sea and sea bed conditions. The company is also engaged in testing equipment at much greater depths in a flooded mine shaft at Rothes Colliery in Fife (Fig. 6.11).

From April 1974, Strongwork Diving International Ltd have been using the loch to test and evaluate the performance of their transfer under-water diving system and to train divers at depth. In December 1974, it was announced by the Manpower Services Commission (p. 63) that a new £2 million diving training centre (the first of its kind for civilian divers in the UK) is to be set up at Fort William. A firm of merchant bankers has been appointed to establish the centre in conjunction with the Training Services Agency. The centre will train men in mixed gas and saturation diving techniques and is expected to become self-financing out of fee-income from the petroleum industry.

The recent (1974) spate of applications for permission to build production platforms on sites on the Clyde estuary and associated sea lochs is in part attributable to changes in platform design and pressures by government to encourage companies to develop sites more readily serviceable by labour from the unemployment 'blackspots' of the West Central Belt (p. 63). However, this may also be considered a result of the difficulties companies experienced in 1973 and 1974 in trying to get permission to develop sites in the North West, notably on the shores of Loch Carron and Loch Broom.

From the engineering standpoint, Loch Carron is reputed to possess the best deep water sites in Scotland for the construction of production platforms of the Condeep gravity type. This encouraged two companies, John Mowlem and Taylor Woodrow (now Taywood Seltrust Offshore), to submit planning applications in the summer of 1973 to develop yards on the same site at Port Cam near Drumbuie (Fig. 6.17). Initially the land required extended to 38 ha (95 ac) and

57 ha (140 ac) respectively, but in November 1973, the companies agreed to amend their proposals to a single jointly operated yard in the hope that this would be more acceptable. This was done soon after the opening of a public inquiry set up by the Secretary of State. All of these plans involved carrying out the earlier stages of platform construction in a dry dock at Drumbuie, and the subsequent floating out of platforms for completion in deeper water, around 150–180 m (500–600 ft) deep, about 9½ km (6 ml) to the west near the Crowlin Islands. It was reckoned that the projects would employ 700–800 men directly. Local and national objections to these proposals and their ultimate rejection by the Secretary of State in August 1974 are discussed in Chapter 7.

Planning applications for three other sites in West Inverness-shire and Wester Ross were submitted in 1973. Of these, one was rejected, one withdrawn and one granted. The application by Taylor Woodrow for a yard on a site at Corran on the north shore of Loch Hourn was rejected by Inverness County Council in June 1973 (p. 113). John Mowlem lodged an application for a yard at Rhue on Loch Broom at the mouth of the Allt an t-Srathain, about 3½ km (2 ml) north of Ullapool, in the summer of 1973. The scheme included the siting of a 10 ha (25 ac) labour camp and storage area on the outskirts of Ullapool, and like the Drumbuie proposals aroused strong opposition locally. Nearly 2 000 objections were lodged with Ross and Cromarty County Council, most of them on grounds of amenity. Matters were resolved in November 1973 when the company withdrew its application following the revelation that the channel between the proposed site and the open sea was too shallow and tortuous for the safe passage of platforms of the design the company intended to build.

In December 1973, the Anglo-French partnership of Howard-Doris (J. R. Howard and Company, London, and C. G. Doris, Paris) lodged a planning application for a concrete platform yard on a 6½ ha (16½ ac) site on the Applecross Estate, about 1½ km (1 ml) south-west of Russel Farm on the west shore of Loch Kishorn. Although the project received approval from Ross and Cromarty County Council in March 1974, despite some 50 objections, no formal approval was given by the Secretary of State until September of the same year.

This development which has a maximum 12-year lease, entails excavating a 137 m (450 ft) diameter dry dock in which the bases of concrete gravity platforms will be built before being floated out into deep water for completion.

Accommodation for a 400-strong labour force

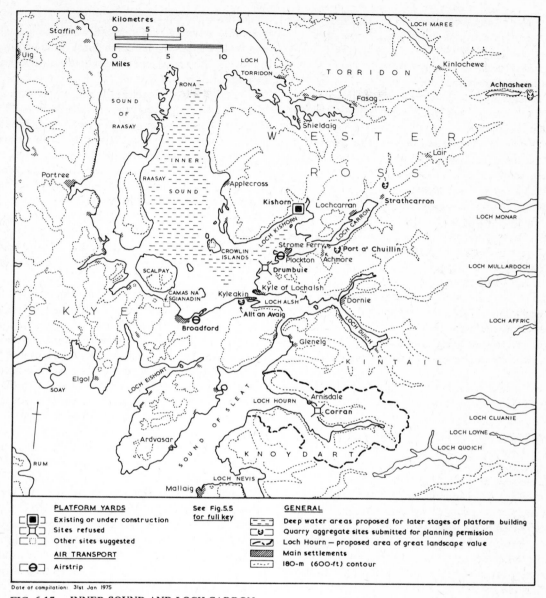

Date of compilation: 31st Jan 1975

FIG. 6.17   INNER SOUND AND LOCH CARRON

is being provided on the site and an access road was completed in December 1974 to link with the existing Applecross-Tornapress road. Because of its remoteness and the need to safeguard local roads and bridges, the site is being treated as an 'island site', that is, equipment and most supplies are brought in by sea. This is one of the stringent conditions laid down for operating the site and further conditions govern the adequate reinstatement of the site once platform production has ceased.

Because of the need to stockpile large quantities

of aggregate on the site (the quantities needed are expected to be worth £2 500 per day once the yard is in production), this material will have to be brought in by sea either from sites in Skye or in the Loch Carron area. By the end of January 1975, Ross-shire had received six applications and Inverness-shire three, to develop aggregate quarries on sites within a twenty-five miles radius of Kishorn (Fig. 6.17). Since most are in areas of considerable landscape value it is not surprising that a number of objections have been received. Whichever sites are given planning approval, they are expected to be subjected to the most stringent conditions concerning both operations and ultimate restoration. At time of writing (January 1975) both counties had deferred making a decision in order to allow joint examination of each of the sites.

In November 1974, Howard-Doris announced receipt of a contract to construct a £60 million concrete gravity platform to serve as the the master unit in the Block 3/3 section of the Ninian Field and further orders are under negotiation. Capital expenditure on the site is expected to amount to £10 million. Of the 76-strong labour force employed in January 1975, all were Scots, mostly people from south-west Ross.

The only other oil-related developments on the north-west mainland remain extremely tentative. Suggestions have been made at various times that Loch Eriboll could be used for deep water wharfage, tank farms for oil storage, a construction yard and a possible oil refinery, and Sutherland County Council have indicated their general interest. However, unless there is a rapid increase in demand by the oil industry for such facilities, the remoteness of the loch seems sufficient to ensure that such proposals are unlikely to be implemented in the near future.

9. Western Isles

An engineering base is currently under construction at Arnish Point near Stornoway for Lewis Offshore Ltd, a subsidiary of the Norwegian company Fred Olsen Ltd. The first stage in development costing £8 million involves facilities for steel fabrication work such as building decks for production platforms. A new quay and stockyard are being built to handle incoming materials (Fig. 6.18). Such work will be done in conjunction with the Aker Group of which Olsen is a subsidiary. A labour force of 500 men rising to 1 000 in 1982 is contemplated. A second stage— the development of a yard for the construction of oil rigs—will depend on the availability of labour.

One of the reasons why development involving employment of this scale can be contemplated on the island is the remarkably high population density of Lewis compared to other parts of the Highlands, including Easter Ross. A further factor is the persistently high unemployment rate (averaging 15% but often exceeding 25% in winter) which has plagued Lewis and Harris for decades. The company have given an undertaking to employ local people and provide training for young workers. It is also hoped that the creation of jobs will

encourage island expatriates to return. Plans for temporary accommodation in Stornoway to house 40 key workers have been passed and a house building programme to provide 250 houses per annum between 1975 and 1979 is being considered.

In the first instance, the project involves a 60 year lease (subject to ten year reviews) of 38 ha (93 ac) of land with options on a further 28 (70 ac). The land has been acquired from the Stornoway Trust Estate which was gifted to the people of Lewis by the first Lord Leverhulme. In addition, a further 5 ha (13 ac) of sea bed for jetties and associated developments have been acquired from the Crown Estate Commissioners. Although the terms of financial settlement have not been revealed, in either case, it is estimated that the deal between the Trust and Lewis Offshore will give the former an assured annual income between £50 000 and £100 000 over the next 60 years.[51]

Another possible future development on the site, should the demand arise, involves the construction of a rig support base. A survey of seven possible sites for service bases in Lewis and Harris which was commissioned by the Highlands and Islands Development Board in 1973 recommended Arnish Point as the most suitable. The other six sites were Stornoway Harbour, Breasclete and Kirkibost (East Loch Roag), Earshader (West Loch Roag), an old whaling station (West Loch Tarbert) and Leverburgh. Although the Stornoway Harbour site had the advantage of existing services and communications, land for development was much more limited than at nearby Arnish Point.

This survey, published in October 1973, followed inquiries by Fred Olsen and two other companies for bases on the island. Approval was given to Fred Olsen Ltd for the Arnish Point site in March 1974 by Lewis Planning Committee, following a study by Ross and Cromarty County Council of the likely impact of such a development on the island. However, negotiations over the acquisition of the necessary land and seabed delayed the start of work on the site until October 1974.

Some concern has been expressed locally over certain aspects of this development, notably its likely effects on existing industries, local life and the Gaelic language. Local industries, notably fishing and fish processing, Harris Tweed and tourism will now face greater competition for labour, and there are fears of the possible disruption of the distinctive island way of life, particularly local values and the survival of Gaelic as a spoken language. The feelings of local people might well be summed up by the Gaelic saying 'is ann oirnn a thàinig an dà làtha' (it's the two days that have come upon us). Having suffered from unemployment and emigration for generations they are faced today with the prospect of welcoming a development which may meet with their material needs but which could completely change their way of life.

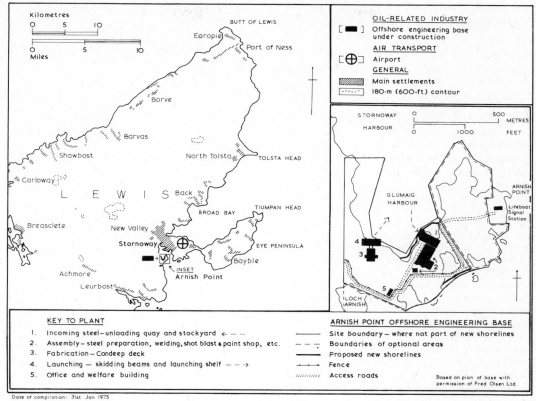

Date of compilation: 31st Jan 1975

FIG 6.18   LEWIS

REFERENCES AND NOTES

1. Harbour Master's Report (1974), Lerwick Harbour Trust.
2. HUTCHESON, A. M. and SMITH, H. D. (eds.) (1973), *Scotland and Oil, Teachers' Bulletin No. 5,* Royal Scottish Geographical Society, Edinburgh, p. 69.
3. Editorial (1974), *The New Shetlander, No. 109* (Yule Number).
4. 74 Eliz. II c. 8, *Zetland County Council Act 1974.*
5. LIVESAY and HENDERSON (1973) *Sullom and Swarbacks Minn Area; Master Development Plan and Report related to Oil Industry Requirements,* Lerwick, Zetland County Council.
6. BAUR, C. (1972), 'Orkney aims to control North Sea oil invasion', *The Scotsman,* 20 September, 1972.
7. 'Flotta folk feel they are being neglected', *The Press and Journal,* 21 November, 1974.
8. 74 Eliz. II c. 30, *Orkney County Council Act 1974.*
9. After spending £2 million on construction works, the Chicago Bridge Company suspended work on this site in August 1974, having failed to get orders for their projected platforms. (Report in *The Press and Journal,* 11 February, 1975).
10. 'MP in bid to boost port of Wick', *The Press and Journal,* 26 November, 1973.
11. SCOTTISH DEVELOPMENT DEPARTMENT (1974), *North Sea Oil and Gas: Pipeline Landfalls – A discussion Paper,* Scottish Development Department, Edinburgh.
12. JACK HOLMES PLANNING GROUP (1968), *The Moray Firth.* Report to the Highlands and Islands Development Board.
13. From a norm of about £200 per acre in the 1960s the price for new leases being negotiated rose to about £2 000 in 1973. Total Marine have paid £3 500 per acre for their (15.2 ac.) site at Altens, and in April 1974 the last (2.4 ac.) site remaining at West Tullos was leased at over £5 800 per acre. (THOMS, A. 'Factory estates "full up" signs on increase,' *The Press and Journal,* 18 February, 1975.)

14. The current crane-loading passenger and cargo vessel on the Aberdeen-Lerwick service is to be replaced by a roll-on/roll-off vehicle ferry when one can be acquired. A new ro-ro vessel, custom-built for the service, was originally ordered from the Hall Russell yard in Aberdeen, but that company relinquished the contract in September 1974. A shortage of tradesmen at the yard, arising from competition for labour with oil-related developments, would have unduly delayed completion of the ship.

15. As many as 30 supply vessels can be in the harbour at one time. Around 70 ship movements per day are averaged in the harbour. Total numbers of supply boat turn-rounds at the harbour have been:

1969	259	*1972*	1 194
1970	335	*1973*	1 720
1971	873	*1974*	2 863 (Provisional figure)

(Information supplied by Aberdeen Harbour Board, January 1975.)

16. Seaforth Maritime was formed in 1972 by a consortium of Scottish shipping, financial and industrial interests to develop opportunities offered by the oil industry. It expanded very rapidly through a series of takeovers and by the formation of joint companies with a diverse range of firms which service or supply oil developments.

17. The Salvesen Group, based in Edinburgh, became in 1973 the first British company to own and operate a drill ship. It has large interests in Aberdeen in fish sales and processing, road transport and ships agency work, as well as a specialist well-casing business.

18. A somewhat similar situation exists in Aberdeen in respect of office developments. Several speculatively built office blocks completed recently in and near the main shopping areas have lain partly vacant while less attractive, but presumably cheaper, units elsewhere have continued to attract tenants.

19. MACKAY, D. I. (1975), *North Sea Oil and the Scottish Economy (North Sea Study Occasional Papers No. 1)*, University of Aberdeen Department of Political Economy, Aberdeen.

20. Among publications which include consideration of the Peterhead area are two published by the Church of Scotland:
    FRANCIS, J. and SWAN, N. (1973), *Scotland in Turmoil*, Church of Scotland Home Board, Edinburgh.
    FRANCIS, J. and SWAN, N. (1974), *Scotland's Pipedream*, Church of Scotland Home Board, Edinburgh.

21. 'Peterhead No to Ammonia', *The Press and Journal*, 11 December, 1974.

22. The Harbours Development (Scotland) Act 1972.

23. Arunta (Scotland) Ltd was set up as an offshoot of a London-based firm of oil industry consultants. In May 1973 the British Oxygen Company (BOC) took a 75% share in the capital of the Scottish company and in November 1974 the company name was changed to Peterhead (BOC) Base Ltd.

24. Methods of calculating the number of vessel turn-rounds vary. Peterhead South Bay (Management) Company provided the following data for vessel entries for seven months July 1974–January 1975. This includes vessels using the BOC and ASCO berths plus those which merely anchored in the bay:

July	478	Oct.	543	Jan.	259
Aug.	447	Nov.	263		
Sept.	495	Dec.	210		

BOC base gave the following as *using* their jetty (figures in brackets give the proportion which are arrivals direct from sea so others are presumably from anchorage in the bay):

July	311	
Oct.	364	(278)
Jan.	176	(113)

ASCO base gave the following as *using* their quay:

July	180
Aug.	164
Jan.	150 (approx.)

25. Peterhead power station will have a capacity of 1 320 megawatts. Its two turbines will be designed to use either oil or natural gas. New 275 kilovolt overhead lines will link it to the grid at Keith and Kintore.

26. A dispute in Aberdeen in 1970 about dues levied on the handling of catches landed there led to a mass withdrawal of inshore fishing boats from that port. Most switched their landings to Peterhead, and within three years the value of white fish landed there rose from about £1 million a year to £9.5 million. A new fish market which was built only in 1971 to help cope with the huge increase in landings is to be doubled in size as part of the current £1 million reconstruction of the North Harbour (Fig. 6.9).

27. Sea Oil Services, a subsidiary company of the P & O group also has offices and warehouse and open storage space in Aberdeen (Fig. 6.8). The Montrose base has been developed at a joint cost to the company and to Montrose Harbour Trust of £6—7 million.

28. In the period June 1974—January 1975 281 merchant vessels visited the port compared with 82 oil supply vessels. In July 1974 the figures were 70 and 14 respectively and in January 1975 30 and 8. (*The Press and Journal*, 6 August 1974 and 4 February 1975).

29. Based on estimates of oil-related employment supplied by Tayside Development Authority.

30. 1 028 ships used Dundee Harbour in 1974, of which 604 were oil-related movements. (*The Press and Journal*, 11 February, 1975).

31. Based on estimates of oil-related employment supplied by Tayside Development Authority and given in 'Arbroath Industry in 1974', *The Arbroath Herald*, 3 January, 1975. It should be noted that a further 100 jobs in sub-contracted engineering and fabrication work were given for the middle of 1974 but only 40 in this category at the end of the year. All figures are approximations and do not compare directly with the Department of Employment figures in Table 5.6.

32. MACKAY, D and MACKAY, A. (1975), 'How decentralisation may prejudice oil development', *The Scotsman*, 5 February, 1975.

33. TRIMBLE, N. (1975), *Estimated Demand for Supply Boat Berths in Scotland, 1974—80* (North Sea Study Occasional Papers No. 2, University of Aberdeen Department of Political Economy, Aberdeen.

34. For instance in the last week of November 1973 the rigs *Staflo, Blue Water 3* and *Sea Quest* were all towed into the Forth for repairs following a severe storm.

35. These Fife anchorages have also been used recently by large tankers temporarily laid up by their owners.

36. SMITH, J. (1974), 'Burntisland oil platform yard plan approved', *The Scotsman*, 27 November, 1974.

37. Granton Harbour has been the base for the construction firms involved, including the operations of a jack-up barge from which piles for the structure were driven.

38. FRAZER, F. (1974a), 'Rivals block SSEB plan to use gas,' *The Scotsman*, 10 April, 1974.

39. FRAZER, F. (1974b), 'BP to build a Forth jetty', *The Scotsman*, 31 August, 1974. HODGE, G. (1974), 'Bright prospects for Forth ports', *The Scotsman*, 9 July, 1974 indicates that 250 000 tons of liquefied gas per year may be exported from Grangemouth.

40. The West Central Belt is taken to include Glasgow, S. Dunbartonshire, Renfrew, N. Ayrshire, N. Lanarkshire and the southern edge of Stirlingshire. 1971 Census population—2.1 million (c.f. Scotland 5.2 million).

41. The most recently available statistics for November 1974 show overall unemployment rates as follows: Glasgow Planning Region 4.6% (Glasgow City 4.9%); Edinburgh Planning Region 3.5%; Tayside Planning Region 3.1%; North East Scotland 2.0% (Aberdeen area 1.4%); Highlands 5.0%

42. Government interest in ensuring the participation of the area in oil-related business has been highlighted by the transfer from London to Glasgow of the headquarters of the Offshore Supplies Office. Glasgow is also a strong contender for the headquarters of the proposed British National Oil Corporation, which if it came to the city could well also encourage oil companies to locate executive offices there.

43. BAUR, C. (1974), 'British Steel moves to slake the insatiable oilfields demand', *The Scotsman*, Oil Register, 23 July, 1974.

44. BAUR, C. (1974), 'Scottish Steel', in *British Steel*, 26, Autumn 1974.

45. Details of firms throughout the UK with interests in developing oil-related business can be found in the *Petroleum Times 1974 Guide to British Offshore Suppliers*, IPC Industrial Press, London.

46. For example, A. Ross Belch, managing director of the Scott-Lithgow shipbuilding group, in evidence to the Portkil public enquiry, 6 January, 1975 and in various other public statements has said that his firm is short of almost 1 200 skilled and semi-skilled men. It lost 440 skilled steel workers between January 1973 and January 1975, of whom about half went to oil-related work.

47. Yet paradoxically its work force has not always been counted in the published totals of such labour (Table 5.6), presumably because only a limited amount of its output could be reckoned to be the result of *North Sea* developments. It was not, for instance, given exemption from 3-day working week restrictions during the fuel crisis of the winter of 1973—74.

48. 'Portavadie inquiry told of sea survey', *The Scotsman*, 27 September, 1974.

49. JACK HOLMES PLANNING GROUP (1974), 'A report to the Government on the physical, economic and social aspects of locating additional platform construction yards round the UK coast.'

50. BAGGOTT, M. (1974), 'Hunterston assurances for Costain Group', *The Scotsman*, 16 May, 1974.

51. 'Lewis "jackpot": Details today', *The Press and Journal*, 16 May, 1974.

7 Politics and Planning

1. Introduction and Outline

Oil-related issues of a political and planning nature necessarily occur and must be considered at a variety of scales from local through regional to national and international levels. From the start of exploration of the offshore geology, decision making has proceeded simultaneously at all four levels, although with considerable shifts of emphasis as the scale of operations has grown.

From the beginning the background to exploration and drilling has been provided by nationally and internationally decided patterns of allocation of licence areas. By contrast, public attention in the earlier stages of development focused mainly on local issues where the oil industry came into conflict with individual communities. Attention has since progressively turned to action being taken at regional and national levels to provide a framework for initiative and control.

In part this change of emphasis simply reflects the increasingly wide distribution of oil-related developments and the growing scale and intensity of effort being put into the search for and exploitation of 'finds'. It also reflects growing awareness by people in areas not immediately affected by the industry of the nature and scale of the opportunities and dangers it presents, both for their regions and for the country as a whole. In addition, it reflects growing realisation of the need to match the multi-national strength of the oil companies and their contractors with something more powerful than ad hoc objectors' committees and the resources normally available to local government. In part it is just an expression of the time taken by politicians and administrators to evolve plans and machinery to cope comprehensively with such developments.

At all times and at all levels the keynote has been one of massive pressure by the oil industry for immediate decisions and action on the schemes it has put forward. Its contractors and suppliers, on the basis of similar pressures put on them by the industry, have been similarly insistent. The repeating theme has been that of the likely losses to companies if decisions are delayed, of their ability to spend vast sums to clear the way for their projects and of their expressed readiness to take their custom elsewhere if early action was not forthcoming. The potential revenues from oil production are so vast that even apparently irresponsibly lavish preparatory expenditure will be recovered several times over—once the oil or gas is brought ashore!

By contrast, the values which local communities have had to set against these pressures have all too often been social, environmental, aesthetic or cultural, rather than more tangible economic ones. It is for this very reason that the struggle against proposed developments has been fought so tenaciously in many cases, but inevitably it has also brought into conflict with the objectors other local people who stand to gain financially from the proposed developments.

This theme is but one of a number which recur at all levels at which political and planning decisions have had to be taken. Others include fears that oil developments will be short-lived and leave behind more problems than they have solved; that the burdens of financing the creation of the more elaborate infrastructure they make necessary will fall unfairly on individual areas or regions; that profits will by-pass those most inconvenienced in the process of producing them; and that politicians and administrators, at whatever level they are operating, are less than masters of the situation because of circumstances obtaining at higher levels over which they have no control. Just how interdependent the various levels are will be indicated in the following discussion, but for the historical reasons given earlier and to demonstrate the structure of decision making, they are examined sequentially.

2. Local Decision Making

At the grass roots level there is no constitutional power for decision making, although it is often at this level that the impact of oil is most acutely felt. In practice, consideration of development proposals lies with town and county councils, major items or disputes being referred to the Secretary of State for decision.

The pressure of events has in many instances given rise to the formation of local organisations to articulate the community response to imminent changes in their environment and way of life. Groups in, for example, Shetland, Wester Ross, the Cromarty Firth area and the Clyde coast have made their influence felt at both regional and national planning levels, in some cases raising substantial sums to further their interests. Likewise, threats to the physical environment have invoked a determined response from conservationists. Naturalists, both amateur and professional, have reacted sharply whenever sites of particular scientific interest have been considered to be threatened (for example the proposal to lay pipes across the Loch of Strathbeg (Fig. 6.6) and site a gas terminal immediately adjacent to it). Rejection of a proposal to build a platform yard on Loch Hourn (p. 106) has been followed by Inverness County Council seeking the approval of the Secretary of State to designate the land round the loch as an area of great landscape value (Fig. 6.17).

If, however, a planning decision made by a local council is of major significance and requires the rezoning of land within the existing development plan for the area, this has to be referred to the Secretary of State for approval. Such proposed changes must be advertised and, if objections are lodged, he must hold a public inquiry to guide him on how to resolve the problem. A similar procedure is followed if a developer appeals against refusal of planning permission by a local authority. Inevitably a public inquiry causes considerable expense and delay in reaching a decision, but the procedure is designed to protect the rights of individuals and communities.

Examples which occurred quite early concerned the projected construction of platforms at Dunnet Bay (p. 75), and of a marine base at Old Torry (p. 85), a nineteenth-century fishing village which had become part of Aberdeen. Here a Town Council decision in October 1971 to replace their previous plans for preserving and modernising over 100 houses by one of demolition raised strong objections. The Town Council's decision was, however, upheld by the Secretary of State, some eighteen months later.

Although decisions to favour or refuse permission for developments may initially be local, final approval or rejection of the plans may thus be removed well outwith the local community. While the Secretary of State must review the arguments with an open mind, his decisions are likely to be influenced by central government policy. The subject of public inquiries is, therefore, returned to in the fourth section of this chapter.

3. Regional Decision Making

In only a few cases were regional plans previously in existence in areas subject to pressures by the oil industry (for example, the Highlands and Islands Development Board's plans for the Cromarty and Moray Firths, p. 76). Under the new system of regional government, however, there is a statutory requirement for each authority to prepare a structure plan and regional report for its area. The new regional councils will also take over the functions of several existing development authorities, such as the North East Scotland Development Authority (NESDA) and Tayside Development Authority (TDA), which were set up in recent years by groups of adjacent county and burgh councils. Such development bodies have been active in attracting oil-related industry to their areas and in encouraging local firms to participate in developments. Bodies such as the Highlands and Islands Development Board and the Scotland West Industrial Promotion Group will continue to exercise their present functions.

In advance of regionalisation, two authorities—Orkney and Zetland County Councils—have taken urgent action to give them greater control over events. This has been accomplished by the promotion of private parliamentary bills specially designed to strengthen their planning and bargaining powers vis-à-vis the oil companies. It is worth examining the Shetland case in some detail because it introduces new dimensions to local government. Further, despite the actions of national government, this example may be followed by other authorities. In this context events in Shetland have also progressed further than elsewhere.

The Zetland County Council Act 1974, promoted originally as a Provisional Order, empowers the Council and its successor, the Shetland Islands Council, to have harbour authority status over two specific areas (Figs. 6.1, 6.3); and further, outwith these specific areas to license works and dredging, and to be created harbour authority in areas surrounding such work; to acquire specific lands; to invest in securities of bodies corporate; and to create a reserve fund. The most important development coming within the sphere of the Act is the Sullom Voe terminal (Fig. 6.3). When completed the terminal will be run by a body called the

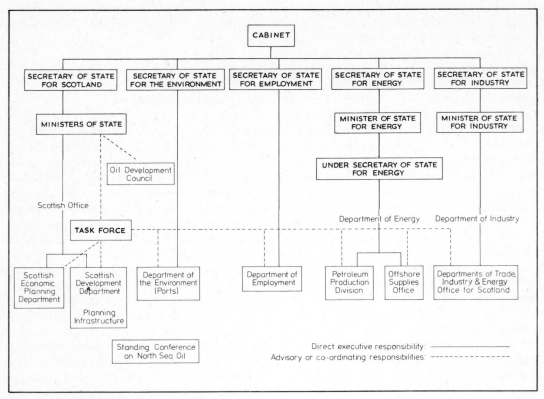

FIG. 7.1   GOVERNMENT ORGANISATION IN RELATION TO OIL

Source:
Reproduced by permission from the *Scottish Office Brief on North Sea Oil*, produced by the Reference Unit, Scottish Information Office.

Sullom Voe Association. Under its constitution, which is being finalised at the time of writing, this body will consist of the Shetland Islands Council and the pipeline operators, that is those oil companies assuming responsibility for supervising the pipelines. It is intended that operating decisions will be taken by the operators only, but all other decisions will require the agreement of the Council.

Thus, apart from being directly involved in commercial undertakings to build the terminal, the Council will be directly involved in what would otherwise have fallen entirely within business control apart from the usual governmental constraints such as planning controls. Any financial surplus accruing to the Council from these commercial undertakings may be applied to public investment for the good of the Shetland community. Further, the instalments of the disturbance payments of the oil companies (p. 73), which could total around £20 million (at 1974 prices) by the end of the century, may also be applied to public investment by the Islands' Council. The Council is already in consultation with central government as to how best these monies should be handled.

4. National Decision Making

At the national scale the key decision on which virtually all others hinge is that setting the rate at which exploitation of offshore resources is to be undertaken. In the UK case the Government decided from the time of the first allocation of licence blocks to encourage companies to undertake exploration as rapidly as possible. Three major factors influenced this decision, and have encouraged subsequent governments to continue the policy:

(i) The desire for economic growth, which makes it necessary to exploit the oil as quickly as possible.

(ii) The beneficial impact which a fast rate of exploitation should have on the UK balance of payments position.

NOTES

Scottish Economic Planning Department (SEPD)
Advises the Secretary of State for Scotland on matters relating to industry and economic development in
Scotland including the development of North Sea Oil, economic resources and Scottish aspects of regional
policies. It is also responsible for general policy in relation to the Secretary of State's responsibilities for
development in the Highlands and Islands, transport, electricity, tourism, new towns and rural development.

Task Force
An inter-departmental committee of senior officials which considers all aspects of oil development which
affect Scotland. Its leader is the Head of the Scottish Economic Planning Department, and its Secretariat
is provided by that Department.

Oil Development Council for Scotland
An advisory body appointed by the Secretary of State for Scotland. It consists of prominent people from many
fields who have knowledge and experience affecting oil development in Scotland. It is chaired by a Minister of
State at the Scottish Office, and its Secretariat is provided by the Scottish Economic Development Department.

Scottish Development Department (SDD)
Has executive responsibility for discharging the statutory functions of the Secretary of State for Scotland in
relation to local government, land use planning and the protection and improvement of the environment, and in
respect of infrastructure services including housing, roads and water supply.

Petroleum Production Division of the Department of Energy
Is concerned with all aspects of licensing policy for oil and gas in the UK Continental Shelf on land. Issues
licences to oil companies and controls all operations under licence; also pipeline policy. Maintains close liaison
with the Offshore Supplies Office.

Offshore Supplies Office, Department of Energy
Headquarters are in Glasgow. Monitors oil companies' purchases. Ensures full and fair opportunities for UK
companies, especially in Scotland, the development areas of England, Wales and Northern Ireland, to compete
for offshore business. Is concerned with promoting new ventures which will increase the UK capacity to serve
the offshore market.

Department of Trade, Industry and Energy Office for Scotland
Representation of headquarters divisions of the Departments of Trade, Industry and Energy in Scotland. Is
concerned with promoting industrial growth and selective financial assistance to industry (including the oil
industry) under the Industry Act; also with industrial location, Department of Industry factories and export
promotion.

Standing Conference on North Sea Oil (SCNSO)
A conference of Local Planning Authorities, Development Authorities, Harbour Authorities, Chambers of
Commerce and the oil companies. The Conference meets twice a year under the chairmanship of a Scottish
Office Minister to exchange views and information.

Ports Office for Scotland, Department of the Environment
Located in Edinburgh. Is responsible for commercial harbours and port developments including those relating
to oil. The head of the office is Chairman of the Scottish Ports Advisory Group (SPAG) which co-ordinates
the activity of all Departments concerned and advises the Task Force on ports matters.

(iii) The strategic significance of North Sea oil for Western Europe, in that it will reduce its dependence upon imports, mainly from the Middle East for long a politically unstable area.

Each of these factors has been reinforced with the passage of time. The North Sea has proved to be a much richer petroleum province than first expected. The vastly increased world price of oil has exacerbated the UK balance of payments position,[1] and with the entry of the UK into the European Economic Community the political interest of Western Europe in North Sea oil has been given added impetus.

The oil companies, as much as the Government, have had a vested interest in exploiting the oil as quickly as possible, both because of their desire to recoup the heavy expenditure involved and because of strong pressure from American interests within many of the companies to meet the increasing US need for imports.

Government functions and responsibilities relative to the oil industry in Scotland are summarised in Fig. 7.1. This framework has been developed since 1970 and the current pattern reflects an increasing accommodation between the government structure and the business needs of the oil industry. Of particular note has been the removal of the headquarters of the Offshore Supplies Office (OSO) from London to Glasgow, and the stated intention of the Government to locate the proposed British National Oil Corporation (BNOC) in Scotland.

One aspect of Government responsibility for oil developments which does not appear in the figure is the concern of the Ministry of Defence for the protection of oil installations in UK waters. Problems of safety and security relating to offshore structures were highlighted in 1974 by the report of a select committee of the House of Commons.[2] Since then the possibilites of protecting offshore operations have been assessed by senior Service officers and plans are expected to be announced soon for the surveillance of rigs and platforms by naval vessels and reconnaissance aircraft. Much stricter limitations on navigation near offshore structures have also been introduced. Meanwhile, the responsibility for policing offshore installations continues to lie with the police force of the particular port from which each is serviced. Thus the Chief Constable of Aberdeen is responsible for rigs located as far away as the East Shetland Basin.

As already indicated (p. 113), many planning decisions on oil-related developments are taken at national level because they involve changes in the zoning of land already approved under local development plans. These decisions are sometimes reached only after the holding of public inquiries, which typically involve considerable delays before schemes are approved or rejected. This tends to conflict with the Government's basic policy to exploit oil as quickly as possible.

Probably the most highly publicised and critical public inquiry to date has been that from November 1973 to April 1974 into the proposals to construct Condeep concrete platforms[3] on a site at Drumbuie, Loch Carron (see also p. 106). This inquiry was of particular interest because it became a test case, not only for developments of the type in question but also for the public inquiry procedure itself. Special attention was attached to the fact that the site involved formed part of the Balmacara Estate bequeathed to the National Trust for Scotland in 1947 'for all time, for the benefit of the nation'. The proposals, therefore, raised the fundamental issue as to whether such areas of 'inalienable' land, gifted under these terms and under strict conservation control, could remain inviolate. If permission had been granted the developers would still have had to go through special parliamentary procedures in order to obtain the site.

Another special feature of the inquiry was the way in which local objectors were backed by national interests, and how this helped to draw attention to the enormous cost of proceedings extending over five months.[44] The involvement of a major national organisa-- tion such as the National Trust for Scotland also introduced people experienced in presenting their case through the mass media, and Drumbuie rapidly became the standard example for journalists with interests in North Sea oil.

Since the local authority had refused permission for development primarily on socio-economic grounds, the proceedings at Drumbuie could be represented in the Press as an attempt by a massive oil-related industry to impose its will on a small rural community. In reality, apart from Drumbuie itself, opinion in the Kyle peninsula was sharply divided. Those in favour tended to view industrialisation as the answer to the more deep-seated problems of the area, particularly as a means of stemming the drift away of the younger elements of the population. Those against argued that the size of the project constituted a serious threat to the existing social pattern and would not in the long term resolve the employment situation.

Technical arguments that the site was uniquely necessary if the UK was to build concrete platforms in competition with foreign yards, and produce them quickly enough to fulfil the Government's timetable for bringing oil ashore, were undermined during the course of the inquiry by external events. Construction of a concrete platform yard at Ardyne Point on the Clyde began in January 1974 (p. 105); Ross-shire County Council approved an application for a similar yard at nearby Kishorn in March 1974 (pp. 106–108); other 'reasonably suitable sites' in the area were distinguished during the course of the inquiry; and certain figures put foward by the Department of Trade and Industry in support of the Drumbuie scheme were shown to have been supplied by the developers and not the result of independent research.

In his rejection of the scheme which came finally in August 1974, the Secretary of State, while accepting that there could be economic advantages to the UK from construction of the Condeep design of platform at Drumbuie, was not persuaded that these were such as to outweigh the loss and burdens which would be imposed upon the locality. He also took account of the inalienable status of the land and the problems this presented.

The Offshore Petroleum Development (Scotland) Bill, which was published by the Government on 7th November 1974, has apparently in part had its origins in the lessons drawn by the Government from the extended nature of the Drumbuie inquiry, together with the greatly increased demand for sites for platform yards in the course of 1973–74 (p. 104). This Bill is far-reaching in its implications. It has aroused considerable controversy and seems likely to be subject to amendment during its passage through Parliament. At present a key provision is that 'The Secretary of State may acquire by agreement or compulsorily any land in Scotland for any purpose relating to exploration for or exploitation of offshore petroleum'.[5] Although the Bill refers to land for any oil-related use it would appear to be particularly aimed at the acquisition of sites for platform construction and pipeline landfalls. An earlier proposal for Government purchase of land urgently needed for these purposes, including a site in the Loch Carron area, was promoted by the Conservative Government in January 1974, but was not followed through due to the change of Government in the following month.[6]

A particularly startling feature of the Bill as tabled was its apparent aim to short-cut by the application of expedited acquisition orders, the procedural safeguards which have previously protected individuals and local communities. As initially designed, the Bill did not make provision for planning procedures to be followed through. At the report stage, however, a Government amendment was approved which makes it explicit that 'no expedited acquisition shall take place unless and until planning permission has been obtained, and unless the land is required for a purpose in accordance with the planning permission'.[7]

In practice the concept of comprehensive planning on a national scale for the location of major oil-related development has evolved only slowly, formulated in a series of discussion papers issued by the Scottish Development Department. The first, issued in April 1973, was a somewhat superficial investigation of the likely demand for production platforms. It distinguished 28 possible sites.[8] In October 1973 an interim oil and gas coastal planning paper suggested that control was needed to avoid over-wide proliferation of oil-related projects. It marked out certain coastal areas as 'preferred conservation zones', and a number of wider areas in east and central Scotland as 'preferred development zones.'

These concepts found fuller definition in a discussion paper on pipeline landfalls, published in May 1974,[9] and in coastal planning guidelines published in August 1974.[10] These form the basis of current Government policy, but as expressions of policy they are very broadly worded, while the map in the most recent publication leaves considerable lengths of coastline undifferentiated. Provision of a more carefully defined planning framework on a national scale would go a long way to achieving a balance between the needs of the oil industry and the long term interests of individuals and communities directly affected.

Reference has already been made to the fact that the main benefits accruing from North Sea oil and gas will lie in the tax revenues they generate—and how such revenues are to be spent—rather than in the direct employment provided by oil-related industry (p. 63). Growing realisation of these facts is a major factor contributing to the steadily growing demand for political change in Scotland in recent years (in part reflected by recent general election results).

In response to increasing political pressures the Government has now recognised the expediency of accepting the idea of devolution in principle, and is promoting the setting-up of an elected Scottish Assembly. While the Government is still showing great reluctance to devolve economic powers, as against purely administrative ones, politicians and economists alike are increasingly recognising that without a much greater measure of economic control than that so far proposed, much of the real responsibility will remain in London.[11]

Additionally, the Government is committed to establishing a Scottish Development Agency (SDA) later this year. Its function is to be the attraction and control of industrial development in Scotland, with a particular concern for the restructuring of the industrial base of the

West Central Belt. A first step in this direction is the transfer of responsibility for industrial development in Scotland from the Department of Industry to the Secretary of State for Scotland. This will become effective on 1st July 1975.

In practice, the task of regenerating the ailing economy of West Central Scotland is likely to be so great that the SDA will need to have access to funds on such a scale as can only be provided by drawing heavily on oil revenues'(p. 63). The decision on how such funds are to be allocated is to remain the prerogative of Westminster, however, so control of the scope of reconstruction of Scottish industry will effectively remain with Central Government. Desire in Scotland for greater control over the wealth from offshore oil and gas is thus unlikely to be satisfied by the arrangements proposed to date. Continuing political pressure on Central Government is likely to continue.

The actual rate at which the Government propose to tax the oil companies has yet (January 1975) to be announced, but it has made clear its intention to impose a Petroleum Revenue Tax (PRT) specially designed to ensure the creaming off of excess profits arising from offshore production. This tax will be imposed in addition to existing arrangements for royalty payments of $12\frac{1}{2}\%$ on the well-head value of oil and normal corporation tax at 52%.

With its added intention of taking a 51% shareholding in all offshore fields, the Government is also seeking to control these developments more directly. At present a sparring match has developed between the Government and the oil companies, as the latter try to influence the decision on the rate at which PRT is to be levied. The politics of oil currently appear fluid, with various companies suggesting how it may easily become uneconomic for them to continue operating in the North Sea. The situation may become clearer in a few months when Government policies are more specifically defined. What remains absolutely clear, however, is that oil must continue to play a major role in both the internal and external politics of the UK. The stakes are high for all concerned. The game must be seen to be played with fairness to all.

REFERENCES AND NOTES

1. In 1974 the deficit on the UK balance of payments on current account was £5 100 million, of which about £3 500 million was accounted for by oil imports. VASSIE, J. D. (1975), 'Pawning Britain's Future,' *The Scotsman*, 4 and 6 February 1975. See also *The Financial Times North Sea Letter*, 13 September 1974.
2. *Report from the Select Committee on Science and Technology: Offshore Engineering, Session 1974, House of Commons Papers, 313,* HMSO, London, 25 July 1974.
3. The Condeep design requires exceptionally deep water for its construction compared with most other concrete gravity platforms.
4. Donations of £26 000 were received by the National Trust for Scotland towards its own costs of £35 000.
5. Clause 1 (1) of the Bill.
6. JAMES, T. (1974), 'Government to buy up production sites', *The Scotsman*, 1 February 1974.
7. JONES, G. (1975), 'Platform sites Bill amended', *The Scotsman*, 15 January 1975.
8. Scottish Development Department (1973), *North Sea Oil Production Platform Towers: Construction Sites—A Discussion Paper*; SDD, Edinburgh, April 1973.
9. Scottish Development Department (1974a), *North Sea Oil and Gas Pipeline Landfalls—A discussion Paper*, SDD, Edinburgh, May 1974.
10. Scottish Development Department (1974b), *North Sea Oil and Gas Coastal Planning Guidelines*, SDD, Edinburgh, August 1974.
11. MACKAY, D. & MACKAY, A. (1975) 'Scotland on Stream — Whose hand is on the oil wealth?' *The Scotsman,* 7 February 1975.

8 Future Prospects

It is now clear that the scale at which oil and gas have been discovered in Scottish waters is much greater than any oil company would have dared to hope only a few years ago. Moreover the same is true of the Norwegian sector, from which much gas and possibly also a considerable amount of oil, will soon be piped to Scottish shores. New discoveries continue to be made almost monthly, while evaluation of known fields quietly adds to the total reserves. It is now apparent that the UK is likely to become a major oil-producing area by world standards, and that Scotland is going to be the source of most of this wealth.

The discoveries, and the onshore activities arising from them, have occurred over such a short period that the economy and administration of the country have hardly had time to make the necessary adjustments. Further, the oil industry has necessarily had to depend initially for many of its requirements on established US and other foreign suppliers, so that home-based industry may seem to have been by-passed to a large extent. There are now growing signs, however, that both the Government and manufacturers are getting past the stage of assessing the opportunities and challenges which the new developments present, and are taking a firmer grip on the situation. It is important that both should tackle their tasks with vision and be prepared to *build on change*, meeting the problems presented with fresh minds and new policies which capitalise on the opportunities presented to the Scottish economy.

Meanwhile, oil has already begun to change both the internal and external geographical relationships of Scotland, and further substantial changes can be expected as the exploitation of offshore wealth develops. Scotland is now becoming an increasingly attractive place for active and ambitious people. Official estimates of population movements for the year to June 1974 show a net gain to Scotland by migration of 7 800, the first such net gain in some 40 years. The Scottish unemployment rate is slowly improving relative to that of the UK, although it still exceeds the latter by a full 1%, and the number of job vacancies is increasing. Much of this overall improvement can be attributed to oil-related expansion in the North East, Easter Ross and Shetland, and a key question over the next few years will be whether this changing balance in the economic geography of the country is to be allowed to continue, or if attempts are to be made to restrain the expanding areas in order to spread the developments more widely.

While there is every need to encourage industry and commerce in West Central Scotland to participate more fully in the oil industry, any policy of diverting developments to it from the new growth areas could work against the interests of the country as a whole. The advantages which the Central Belt could gain in terms of employment would be small in relation to its total need, and the advantages which oil-related businesses seek in locating close to one another would be lost (pp. 63 & 86).

Externally, new trading patterns and political relationships are evolving both within the UK and more widely, between Scotland and Europe and between Scotland and North America. Such trends seem bound to continue over the next decade at least, as oil production from Scottish waters expands towards its peak.

The new situation in which Scotland finds itself requires willingness to accept new ideas and changed relationships. Opportunities are on offer for those who are prepared to see them and to take the initiative. Oil has opened new horizons for Scotland in many directions. The next decade will test the ability of the Scots to reach out towards them.

Bibliography

Oil Industry General

British Petroleum Company Ltd. (1970), *Our Industry Petroleum*, London
British Petroleum Company Ltd. (1973), *BP Statistical Review of the World Oil Industry—1973,* London.
Cazenove & Co. (1974), *The North Sea: the search for oil and gas and the implications for investment*, London.
FOX, A. F. (1964), *The World of Oil*, Pergamon, Oxford.
Investors Chronicle/Petroleum Times (1974), *North Sea Oil Report*, 3 May 1974. [Supplement issued with these periodicals]
ODELL, P. R. (1965), *An Economic Geography of Oil*, Bell, London.
ODELL, P. R. (1974), *Oil and World Power*, 3rd Edition, Pelican Books, London.
Oil Committee, The Organisation for Economic Co-operation and Development (1973), *Oil—The Present Situation and Future Prospects*, OECD, Paris.

Geology and Engineering

ALLCOCK, L. C. (1974), 'Offshore oil and gas: the future engineering contribution', *Instit. Mech. Engrs. Proc* 188 (2).
Bank of Scotland Oil Division (1974), *Facts about Oil,* Bank of Scotland, Edinburgh.
DELANY, E. M. (ed.), (1971), *The geology of the East Atlantic continental margin 2: Europe, Inst. Geol. Sci. Report 70/14,* HMSO, London.
Department of Energy, (1974), *Guidance on the Design and Construction of Offshore Installations,* HMSO, London.
GOLDBERG, E. G. (ed.), (1973), *North Sea Science*, MIT Press, Cambridge, Mass.
HINDE, P. (1974). *The exploration for petroleum with particular reference to North West Europe,* British Gas Corporation, London.
Institute of Petroleum *et al.* (1974), *Petroleum and the continental shelf of North West Europe: the geology and the environment.* [Abstracts from papers read at a conference in London, November 1974].
New Civil Engineer Special Review: North Sea Oil, May 1974.
New Civil Engineer Special Supplement: Offshore Structures, September 1974.
Offshore Installations: The Offshore Installations (Construction and Surveys) Regulations 1974, Statutory Instrument No. 289, HMSO, London.
SISSONS, J. B. (1967), *The Evolution of Scotland's Scenery,* Oliver and Boyd, Edinburgh.
STEERS, J. A. (1973), *The Coastline of Scotland,* Cambridge University Press, Cambridge.
TROUP, K. D. (ed.), (1973), *The North Sea Spectrum*, Thomas Reed, London.
Warren Spring Laboratory, Department of Trade and Industry (1972), *Oil pollution of the sea and shore: a study of remedial measures,* HMSO, London.

United Kingdom

Continental Shelf (Jurisdiction) Order 1968, Statutory Instrument No. 892, HMSO, London.
Department of Energy, (1974a), *Production and reserves of oil and gas in the United Kingdom: a report to Parliament*, HMSO, London, May 1974.
Department of Energy, (1974b), *United Kingdom offshore oil and gas policy: a report to Parliament*, HMSO, London, July 1974.
Department of Trade and Industry, (1973), *North Sea Oil and Gas: a report to Parliament*, HMSO, London [Reprinted from *Trade and Industry*, **10**, No. 4, 25 January 1973].
Heriot-Watt University, (1973), *North Sea oil—the challenge and the implications,* Heriot-Watt University Lectures 1973, Edinburgh.
International Management and Engineering Group (IMEG), (1973), *Study of potential benefits to British industry from offshore oil and gas developments,* HMSO, London.
Report from the Select Committee on Science and Technology: Offshore Engineering, Session 1974, House of Commons Papers, 313, HMSO, London, 25 July 1974.

Scotland

DEKKER, N. (1973), *The Reality of Scotland's Oil, Upthrust Series No. 2*, SNP Publications, West Calder, Midlothian.

FRANCIS, J. & SWAN, N. (1973), *Scotland in Turmoil*, Church of Scotland Home Board, Edinburgh.

HUTCHESON, A. M. & SMITH, H. D. (eds.), (1973), *Scotland and Oil, Teachers' Bulletin No. 5*, Royal Scottish Geographical Society, Edinburgh.

MACKAY, D. I. (1975), *North Sea Oil and the Scottish Economy, North Sea Study Occasional Papers No. 1*, University of Aberdeen Department of Political Economy, January 1975.

North Sea Oil Information Sheet: February 1975, NSOIS 75(1), North Sea Oil Division, Scottish Economic Planning Department, Edinburgh, 20 February 1975. [First of a new series.]

Oil Development Council for Scotland, Committee on the Environment, (1974), *North Sea Oil and the Environment*, HMSO, London.

Scottish Council (Development and Industry), (1972), *Oil and Scotland's Future*, Papers read at the Second International Forum, Aviemore, 24–25 February 1972, SC(D and I), Edinburgh.

Scottish Council (Development and Industry), (1973), *A Future for Scotland*, SC(D and I), Edinburgh.

Scottish Council (Development and Industry), (1974a), *United Kingdom Offshore Oil and Gas: An Assessment of the Expected Production from Existing Finds in Scottish Waters*, SC(D and I), Edinburgh, April 1974.

Scottish Council (Development and Industry), (1974b), *Investing in Scotland's Future*, Papers read at the Fifth International Forum, Aviemore, 6–7 November 1974. SC(D and I), Edinburgh.

Scottish Development Department (1973), *North Sea Oil Production Platform Towers: Construction Sites–A Discussion Paper.* SDD, Edinburgh, April 1973.

Scottish Development Department (1974a), *North Sea Oil Pipeline Landfalls–A Discussion Paper*, SDD, Edinburgh, May 1974.

Scottish Development Department (1974b), *North Sea Oil and Gas Coastal Planning Guidelines*, SDD, Edinburgh, August 1974.

Scottish Education Department (1974), *North Sea Oil and Further Education*, Scottish Information Office, Edinburgh.

Scottish Office, (1973), *Scottish Economic Bulletin: Special Number–North Sea Oil*, HMSO, Edinburgh.

Standing Conference on North Sea Oil–Information Sheets, April 1972–November 1974, North Sea Oil Division, Scottish Economic Planning Department, Edinburgh. [This series has been replaced by North Sea Oil Information Sheets produced by the same department (see above).]

TRIMBLE, N. (1975), *Estimated Demand for Supply Boat Berths in Scotland, 1974–80, North Sea Study Occasional Papers No. 2*, University of Aberdeen Department of Political Economy, January 1975.

Regional

B. P. Refinery (Grangemouth) Ltd (1974), *The Forties oil field and the Forth Valley*, British Petroleum, Grangemouth.

FRANCIS, J. & SWAN, N. (1974), *Scotland's Pipedream*, Church of Scotland Home Board, Edinburgh. [Peterhead area.]

North East Scotland Development Authority (1974), *North East Scotland and the Offshore Oil Industry: A summary of the Main Developments, 1974–No. 2*, NESDA, Aberdeen.

Scottish Development Department (1974), *Examination of Sites for Gravity Platform Construction on the Clyde Estuary*, SDD, Edinburgh.

Zetland County Council (1974), *Sullom Voe District Plan*, ZCC, Lerwick.

Directories

Financial Times Business Enterprises Division (1974), *'Financial Times–Who's Who in Oil and Gas.' Financial Times*, London.

GRAHAM, P. *et al.* (1974), *United Kingdom Offshore Oil and Gas Yearbook*, Kogan Page Ltd., London.

Highlands and Islands Development Board, (1974a), *A Directory of Industry, Commerce and Public Administration in the Highlands and Islands*, HIDB, Inverness.

Highlands and Islands Development Board (1974b), *Offshore Directory*, HIDB, Inverness.

Institute of Offshore Engineering, Heriot-Watt University, (1973), *Guide to Information Services in Marine Technology*, Institute of Offshore Engineering, Heriot-Watt Univeristy, Edinburgh.

MACAULAY, A. J. (1974), *North Sea Oil 1973, Select List of Periodicals and Newspaper Articles*, Department of Library and Business Information, Edinburgh College of Commerce.

North East Scotland Development Authority (1974), *North East Scotland and the Offshore Oil Industry: A Summary of the Main Developments, 1974–No. 2*, NESDA, Aberdeen.

North of England Development Council (in co-operation with the Offshore Supplies Office), (1974), *Oilfield*, 2nd Edition, NEDC and OSO, Newcastle upon Tyne.

Offshore Supplies Office (1974), Untitled information folder for firms interested in offshore developments, OSO, Glasgow.

Petroleum Times (in co-operation with the OSO), (1974), *1974 Guide to British Offshore Suppliers*, IPC Industrial Press, London.

Reference Unit, Scottish Information Office (1974), *Scottish Office Brief on North Sea Oil*, Scottish Information Office, Edinburgh.

SHEPHERD, P. (1974), *Oil: A bibliography,* 2nd Edition, Aberdeen and North of Scotland Library and Information Co-operative
Service (Anslics), Aberdeen.
SHEPHERD, P. (1975), *Oil: A bibliography,* 2nd Edition, 1st Supplement, Anslics, Aberdeen. [Bibliographies on oil and oil-related
subjects held in libraries in North East and North Scotland.]

Periodicals

Business Scotland, (Monthly), Holmes McDougall, Edinburgh. [Especially Oilscene section.]
Construction News, (Weekly), Construction News Pub. Ltd., London.
Financial Times North Sea Letter, (Fortnightly), *Financial Times*, London.
New Civil Engineer, (Weekly), *New Civil Engineer*, London.
Noroil, (Monthly), Noroil Publishing House, Stavanger. (Also Summer Street, Aberdeen.)
Offshore, (Monthly), Petroleum Publishing Company, London.
Offshore Engineer, (Monthly), *New Civil Engineer,* London.
Offshore Services, (Monthly), Spearhead Publications Ltd., Kingston-upon-Thames.
The Oilman, (Weekly), Maclean-Hunter Ltd., London.
The Petroleum Economist, (Monthly), *Petroleum Economist*, London.
Petroleum International, (Monthly), Petroleum Publishing Company, London.
Petroleum Times, (Fortnightly), IPC Industrial Press Ltd., London.
Roustabout, (Fortnightly), Roustabout Publications, Aberdeen.

Newspapers

Newspapers are a useful source of information on the rapidly changing oil scene. Special supplements are regularly produced, of which *The Scotsman Oil Register* is perhaps the most valuable in the Scottish context. In addition *The Times, The Scotsman, The Glasgow Herald, The Press and Journal* and *The Courier* provide good daily coverage of the oil industry. Local weeklies, such as *The Shetland Times* and *The Orcadian*, also provide a valuable service in this context.

Glossary

1. Terms describing Petroleum

Petroleum is the collective term which describes the whole range of solid (e.g. bitumens), liquid and gaseous compounds of carbon and hydrogen which are naturally formed and sealed up in the rocks of the earth's crust. It consists mainly of crude oil and natural gas in infinitely variable proportions.

Crude Oil is the oil produced from an underground reservoir after being freed of any gas which may have been dissolved in it under reservoir conditions, but before any other operation has been performed on it.

Natural Gas is gas found in the earth's crust under pressure and often produced in association with crude petroleum, where its pressure helps in the recovery of the latter in many cases.

Gas Condensate or *Condensate* is a mixture of gas and light coloured crude oil.

2. Measurement of Petroleum

Billion: Throughout the text 1 billion = 1×10^{12} (as used by BP).

Tonne: Refers to a metric ton. 1 tonne = 1 000 kg. (2 204.6 lb).

Barrel: Crude oil is measured in barrels. 1 barrel = 35 Imperial gallons.
Rate of production is usually expressed in barrels per day (b/d).
Although the factor for converting tonnes to barrels depends on the specific gravity of the crude oil, a useful approximate conversion of tonnes to barrels may be obtained by multiplying by 7.5.
Also an approximate conversion of the rate of production may be made as follows: 1 barrel per day = 50 tonnes per year.

Natural Gas: Approximate calorific equivalents are:
   1 therm $\equiv$ 3.05 m^3 (107.53 ft^3)
   1 tonne of crude oil $\equiv$ 1 133.54 m^3 (40 000 ft^3) of natural gas.

Reserves: *Proven*: The quantity of oil or gas discovered and proved by drilling.
   Recoverable: The quantity of oil and gas which might be recovered by available methods.

3. Offshore Engineering

Rig is commonly used to cover not only the derrick and drilling equipment, but also the ship or other structure on which these are mounted.

Jack-up rig is a floating deck with legs passing through it which can be lowered to rest on the sea bed (Fig. 3.9).

Semi-submersible rig is a deck supported by legs attached to submerged pontoons which by water ballasting can be sufficiently stabilised to permit drilling in conditions of wave motion of the kind encountered in the open sea

(Fig. 3.9). These rigs are probably more suitable for use in the deeper northern North Sea, especially in winter.

Drill-ship is a specially built (or converted) ship complete with derrick and drilling equipment operated through its bottom.

Production platform is a large steel and/or concrete structure fixed to or resting on the sea bed respectively, for the extraction of offshore petroleum. There are several main types:

> *Jacket (steel) structures* are braced steel frames secured to the sea bed by means of steel piles and with no capacity for the storage of oil (Fig. 3.7)

> *Concrete gravity structures* are reinforced or pre-stressed concrete structures relying on gravity to resist overturning forces from waves and other environmental loadings, and having provision for storing oil (Fig. 3.7).

> *Hybrid (composite) structures* are hollow concrete tanks surmounted by steel towers to support the working platform. The base gives stability by means of gravity action and provides oil storage.

> *Tension-leg structures* each consist of a deck attached to a submerged buoyancy chamber which is in turn anchored to the sea bed by a series of wires (composed of steel or other materials). The tension of the wire is maintained by the floating buoyancy chamber (Fig. 3.7).

Deck is the top of a rig or platform upon which the derrick and modules are fixed.

Jacket is the braced steel frame of a steel production platform.

Node is a point of junction of a number of tubular sections in a steel platform jacket. This a key stress area in the structure.

Module consists of the steel superstructure for housing living quarters, generators, laboratories and other items on the deck of a rig or production platform.

Subsea production system is a method of extraction involving the enclosure of well heads and related production equipment in sealed modules placed on the sea bed. A *subsea completion* is a well head producing oil linked to a platform, buoy or SPAR, rather than directly to the platform via a marine riser.

ELSBM (Exposed location single buoy mooring—Fig. 3.9) and *SPM* (Single point mooring—Fig. 6.4) are mooring systems designed to minimise the influence of environmental conditions during tanker loading by allowing tankers to swing round a fixed buoy or tower in response to wind, wave and tidal action.

SPAR is a similar mooring system in which the buoy also forms a storage unit. (Fig. 3.9).

4. Drilling

Blow-out preventer is a special hydraulically operated gland-like device, employing synthetic rubber, designed for use while drilling, to maintain pressure control of the drilling fluid.

Derrick consists of a steel pylon-like structure which with ancillary equipment is used both to support and raise and lower other drilling equipment and aid in the carrying out of other operations connected with an oil well.

Drill bit is the rotating piece of machinery carrying cutting edges and attached to the bottom of the drill string.

Drill pipe is the steel pipe used for carrying and rotating the drilling tools and for permitting the circulation of the lubricating mud.

Drill string is the series of drill pipes connecting the drill bit to the derrick machinery.

Drilling mud consists of a combination of water and chemicals, notably barytes, used to lubricate the rotating drilling equipment and to carry rock cuttings to the surface of the well.

Marine riser is a vertical pipe, connecting the blow-out preventer to the rig or platform, through which the drilling takes place and the circulation of drilling mud is maintained.

Well casing is the steel lining of a well, the main purposes of which are to prevent the caving in of the sides, to exclude water or gas, and to provide means for the control of well pressures and crude oil production. The lining is cemented in place and, like drill pipe, is 'run in strings'.

Well logging consists of the operation of sampling rock cuttings taken to the surface, otherwise recording the nature of the rock strata encountered, and preparing a record of the geological column in a well.

Step out wells are a series of wells drilled in order to appraise a discovery after drilling of the first exploration well.

Wildcat well is a well which has been sunk without a complete geological exploration of the locality.

5. Economic

Posted price is the price set on oil when it crosses the border of a producing country.

Royalty is a tax levied by governments on the well-head value of petroleum produced.

Well-head value is derived by deducting from the landed value reasonable costs incurred in delivering the oil and gas ashore.

6. Conversion Factors

	Metric Unit	Imperial Equivalent	
		Accurate	Rough approximation
Length	1 metre (m)	3.281 ft	10 m = 33 ft
	1 kilometre (km)	0.621 ml	8 km = 5 ml
			100 km = 62 ml
Area	1 square metre (m²)	10.764 ft²	11 ft²
	1 hectare (ha) (10 000 m²)	2.471 ac	2.5 ac
Volume	1 cubic metre (m³)	35.288 ft³	35 ft³

Note: On maps the 600 ft contour has been shown as 180 m.

Postscript

Significant developments between 1 February and 30 June 1975 include:

1. Steadily increasing operating and development costs — e.g. operating costs of a large semi-submersible exploration rig are now put at as much as £30 000 per day (c.f. p. 16). Development costs of the Forties Field are now estimated at £700 million (c.f. p. 48).

2. The 1 000 ft dive (c.f. p. 22) — The world's deepest commercial saturation dive was achieved in June 1975 off Labrador when two teams of divers recovered a blow-out preventer stack (Glossary) from its site on the sea bed at 1050 ft (320 m).

3. Oil production — The first oil to be produced commercially from a Scottish offshore field (Argyll) was landed by tanker at BP's Isle of Grain refinery in the Thames estuary on 18 June 1975. Production is being undertaken using a converted semi-submersible rig as a temporary platform (Table 5.5). Tankers loaded at the field by the SPAR method (Fig. 3.9) will transport the oil to several UK refineries.

4. Forties Field production (c.f. p. 20 Table 5.5) — All four production platforms were on location by the end of June and drilling of the first production well began from Greythorp I on 22 June 1975. The first oil is now expected to be piped ashore in September or October 1975.

5. *Oil Industry Employment as at 30 April 1975* (Addition to Table 5.6, p. 61)

Inverness and Easter Ross	5 195[a]
Remainder of Highlands and Islands	2 350[b]
North East	7 635
Tayside	755
East Central	1 970
West Central	1 010
Total	18 915[c]

[a] Increased recruitment in construction yards at Nigg and Ardersier to meet float-out dates for major structures within the 1975 'weather window'.[1]

[b] Construction work in progress at Sullom Voe, Kishorn and Portavadie; also expansion of work force at Ardyne Point.
[c] An overall increase of 16.5% in four months in the total work force employed on projects established directly as a result of North Sea oil developments.

6. Business failure of floating engineering workshop (c.f. p. 78) — This venture, undertaken by a consortium of shipping, shipbuilding and financial interests — Marine Oil Industry Repairs (MOIRA) — with financial support from the Department of Energy, was declared bankrupt in mid-June and the entire labour force was paid off by the end of the month. The workshop had failed to attract a major contract during the three months it had been on station in the Cromarty Firth.

7. Portavadie platform yard (c.f. Fig. 6.16 and p. 105) — Development of this site is now well advanced, but no orders for it have been announced so far. Substantial Government financial backing has been provided for the project with a view to the yard being ready to produce its first platform to meet a delivery date within the 1977 'weather window'. Such a delivery target was, however, dependent upon early receipt of a contract and no orders at all for concrete platforms have been placed with UK yards since development of Portavadie began in February.

8. Petroleum and Submarine Pipelines Bill (c.f. pp. 31 and 114) — Still under consideration by Parliament at the end of June, but the Government has made clear its intention to establish the British National Corporation this year, with headquarters in Glasgow.

9. Oil Taxation Bill (c.f. p. 118) — Has received Royal Assent. Rate of Petroleum Revenue Tax (PRT) has been fixed at 45%. In addition, petroleum production will continue to be subject to Royalty at 12.5% of well-head value and to Corporation Tax on net revenue at 52%. PRT will be charged on a field-by-field basis.
   Allowances include:
(i)   175% allowance on field development expenditure.

[1]   The term 'weather window' is applied to that part of the year when major offshore operations such as pipelaying and the floating out and positioning of large structures is less likely to be interrupted by high winds or heavy seas.

126

(ii) Production of a maximum of 500 000 tonnes will be allowed free of PRT each half-year, subject to a cumulative total of 10 million tonnes per field.

(iii) Exemption from PRT on gas contracted for sale to the British Gas Corporation before 30 June 1975.

(iv) Cancellation of PRT where it would reduce pre-Corporation Tax profit to less than 30% of the investment cost to date.

Further, where development of marginal fields is considered to be in the national interest powers are being sought through the Petroleum Bill to refund Royalties in whole or in part.